Springer Proceedings in Earth and Environmental Sciences

Series Editor

Natalia S. Bezaeva, The Moscow Area, Russia

The series Springer Proceedings in Earth and Environmental Sciences publishes proceedings from scholarly meetings and workshops on all topics related to Environmental and Earth Sciences and related sciences. This series constitutes a comprehensive up-to-date source of reference on a field or subfield of relevance in Earth and Environmental Sciences. In addition to an overall evaluation of the interest, scientific quality, and timeliness of each proposal at the hands of the publisher, individual contributions are all refereed to the high quality standards of leading journals in the field. Thus, this series provides the research community with well-edited, authoritative reports on developments in the most exciting areas of environmental sciences, earth sciences and related fields.

More information about this series at http://www.springer.com/series/16067

Rosemary M. Gutierrez
Editor

Wastewater Technologies and Environmental Treatment

Proceedings of the ICWTET2020

Springer

Editor
Rosemary M. Gutierrez
Department of Biology
University of the Philippines Baguio
Baguio City, Philippines

ISSN 2524-342X ISSN 2524-3438 (electronic)
Springer Proceedings in Earth and Environmental Sciences
ISBN 978-3-030-61988-6 ISBN 978-3-030-61989-3 (eBook)
https://doi.org/10.1007/978-3-030-61989-3

This Springer imprint is published by the registered company Springer Nature Switzerland AG
The registered company address is: Gewerbestrasse 11, 6330 Cham, Switzerland

Preface

Dear Distinguished Authors and Guests,

It was a great pleasure to welcome all participants in the 2020 the 2nd International Conference on Wastewater Technologies and Environmental Treatment (ICWTET2020). Due to COVID-19 pandemic which is currently affecting many countries, ICWTET2020 had been held on June 19–20, 2020, as a webinar.

Its purpose is to serve as an international forum for the presentation and discussion of recent advances in the field of wastewater technologies and environmental treatment. ICWTET facilitated the sharing of data and ideas, promote collaborations, and address common challenges to wastewater technologies and environmental treatment. ICWTET provided an effective environment for active learning, discussions of real-world case studies, and networking between like-minded peers and scientists with interest in the field of wastewater technologies and environmental treatment.

I wish to thank the authors and the reviewers for contributing to ICWTET2020. I also express my gratitude to all technology committee members, conference chairs, keynote speakers, sponsors, and conference participants for their support and contributions to ICWTET2020. With great hope, I wish the organizers a most successful conference in ICWTET2021.

I look forward to your participation in ICWTET2021.

With our warmest regards

Prof. Rosemary M. Gutierrez
University of the Philippines Baguio
Baguio City, Philippines

Contents

Effects of Argillaceous Shale Efflorescence on Soil Water Content and Soil Properties in Newly-Increased Farmland

Qiguang Dong, Na Li, Yufei Xiong, and Jing He

Abstract In the land remediation of the Weibei Loess Plateau, a large number of argillaceous shale efflorescence was mixed with newly-increased farmland soil. To study the effect on soil water content in newly-increased farmland by argillaceous shale efflorescence, a mixture of argillaceous shale efflorescence and local loess, ancient soil was researched in this study to investigate its effect on soil water in two types of soil. The results showed that the addition of argillaceous shale efflorescence could slow the infiltration of loessial soil to a certain extent and improve the water holding performance; while the argillaceous shale efflorescence could accelerate the infiltration of soil water to a certain extent after adding to the clayey ancient soil. The argillaceous shale could also improve the quality of the newly cultivated land and has potential application value for improving the soil texture.

Keywords Argillaceous shale efflorescence · Soil water content · Newly-increased farmland

Q. Dong (✉) · Y. Xiong · J. He
Institute of Land Engineering and Technology, Shaanxi Provincial Land Engineering Construction Group Co. Ltd., Xi'an, China
e-mail: dq-guang@163.com

Q. Dong · N. Li · Y. Xiong · J. He
Shaanxi Provincial Land Engineering Construction Group Co. Ltd., Xi'an, China

Q. Dong · Y. Xiong · J. He
Key Laboratory of Degraded and Unused Land Consolidation Engineering, The Ministry of Natural Resources, Xi'an, China

Q. Dong · Y. Xiong · J. He
Shaanxi Provincial Land Consolidation Engineering Technology Research Center, Xi'an, China

R. G. Gutierrez (ed.), *Wastewater Technologies and Environmental Treatment*, Springer Proceedings in Earth and Environmental Sciences,
https://doi.org/10.1007/978-3-030-61989-3_1

1 Introduction

Land remediation is the management of inefficient, unreasonable and unused land, and the restoration and utilization of land damaged by production and construction and natural disasters to improve land utilization, including agricultural land consolidation, land development, land Reclamation, construction land improvement, etc. (Yang et al. 2013; Liu et al. 2015), which is an important way and effective measure to increase the area and quality of cultivated land. The current land improvement project has been implemented for many years, and the amount of new cultivated land has increased considerably, but its quality is not optimistic (Yang 2011).

In the Weibei Loess Plateau of Shaanxi Province in China, there are abundant wasteland resources, and most of them have not been used reasonably. Implementing land improvement projects in this area can effectively achieve the balance of occupation and compensation of cultivated land resources in Shaanxi Province. Soil quality is the main criterion for judging whether new cultivated land can achieve high yield and efficiency and whether land remediation can achieve sustainability (Zhou et al. 2014). Due to soil formation and the impact of human activities, many soils in the Weibei Loess Plateau contain various amounts and types of gravel or other rock efflorescence (Zhou et al. 2011). During the process of rehabilitating desert grasslands on the Loess Plateau, it was found that a large number of argillaceous shales existed in some areas, and its efflorescence was often mixed into the newly cultivated soil. This has changed the physical structure of the soil and the stability of the soil, and caused changes in the way of water movement in the soil, which affected the redistribution of water (Mehuys et al. 1975; Xu and Shang 2008; Wang et al. 2012; Yang et al. 2009).

At present, there are many studies on the improvement of soil quality of new cultivated soil improvement additives. According to the source of raw materials, soil improvers can be divided into natural improvers, synthetic improvers, natural-synthetic copolymer improvers, and biological improvers (Chen and Dong 2008). Among them, natural modifiers are relatively low in cost and easy to obtain, and are mainly divided into inorganic materials and organic materials. Relevant research on the effects of inorganic materials on soil quality has mostly focused on zeolites, bentonite, fly ash, gypsum, vermiculite and other materials (Hao et al. 2005; Wang et al. 2005). Zeolite has a good water storage capacity. After being applied to the soil, it can increase the water content of the cultivated layer of the soil by 1–2%, increase the field water capacity of the cultivated layer of the soil in the arid area by 5–15%, and its strong adsorption The capacity and high cation exchange capacity can adsorb heavy metals in the soil, promote the release of nutrients in the soil, and have certain effects for improving the status of water and fertilizer shortages in newly-added cultivated soils. Using bentonite to improve sandy soil can increase the content of clay and increase the moisture content of the soil. In addition, bentonite has a certain swelling, dispersibility, and adhesiveness. Applying soil can increase the number of aggregates, increase soil porosity, and reduce soil. Bulk density etc. Fly ash, as an inorganic solid waste, is also commonly used to improve the physical properties of clay and sandy soils, and can increase the content of boron, zinc, and silicon in the soil. Although these natural minerals have certain effects in improving soil structure and soil chemical properties, most of the related studies are performed on a single type of soil, and most of them are focused on the improvement of sandy soils, and very few Considering the characteristics of soils with different texture types, there are still some theoretical and technical problems in practical applications, such as the application amount, application method, and the limitation of large-scale popularization and application of these natural mineral reserves. To this end, exploring the universal addition of relatively abundant externally added materials and studying its impact on the soil quality of newly cultivated land with different texture types is of great significance to further improve the quality of newly cultivated land and promote the ecological construction of land improvement projects.

Based on the physical and chemical characteristics of argillaceous shale, this study intends to set up a combination of argillaceous shale efflorescence with local loess and ancient soil, and discuss its impact on the soil water distribution of newly-increased farmland.

2 Materials and Methods

The study selected mud shale efflorescence and different types of soil for mixing and mixing, and set up a pot experiment. The argillaceous shale efflorescence and soil samples selected for the study were collected from a land improvement project area in Tongchuan City, Shaanxi Province. Among them, the soil samples are two types of loess soil and paleo soil. The texture of the soil is silty loam. The texture of the collected loessal soil is powdery loam. The collected samples are air-dried and passed through a 2 cm sieve. The shale efflorescence materials are mixed with loessal soil and ancient soil in different proportions. Alfalfa is planted as test plants. The specific test treatment is shown in Table 1. For each treatment, a quantitative irrigation was performed on April 17, 2019, and the soil water content of the mixed soil was periodically measured after the irrigation. The soil water content was sampled by a ring knife and measured by the drying method. The measurement depths were 0–20 cm and 20–40 cm. Each measurement was repeated 3 times.

Table 1 Treatment of the study

Test treatment	Vs (argillaceous shale):Vl (loessal soil)	Test treatment	Vs (argillaceous shale):Vp (paleosol)
H1	1:1	G1	1:1
H2	1:2	G2	1:2
H3	1:5	G3	1:5
CK1	0:1	CK2	0:1

3 Results and Analysis

3.1 Influence of Argillaceous Shale Efflorescence on Loess Soil Water Content

The soil water content is affected by water replenishment and the characteristics of the soil itself. Figures 1 and 2 reflect changes in the average value of soil water content at different depths after mixing loessal soil and argillaceous shale in different proportions. It can be seen from Fig. 1 that in the 0–20 cm surface soil, CK1 (Vs:Vl = 0:1) and H3 (Vs:Vl = 1:5) show a trend of first increase and then decrease (Vs:Vl = 1:1) and H2 (Vs:Vl = 1:2) both show a trend of increasing first and then stabilizing. Among them, CK1 decreased from 44.63 to 31.98%, a decrease of 12.65%, and H3 decreased from 38.85 to 26.07%, a decrease of 12.78%. The soil

Fig. 1 Variation of soil water content of 0–20 cm in mixed soil of loessal soil and argillaceous shale

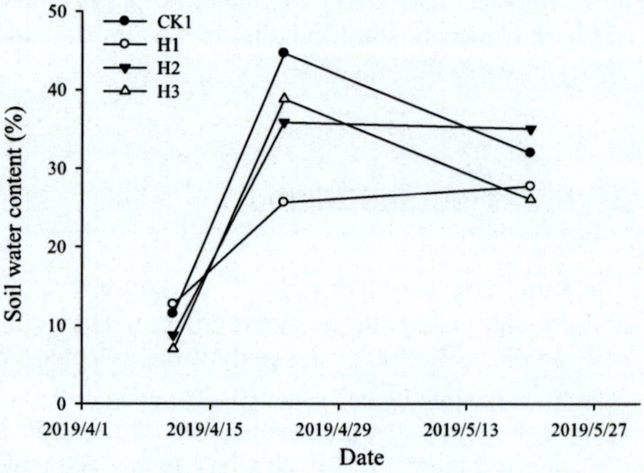

Fig. 2 Variation of soil water content of 20–40 cm in mixed soil of loess soil and argillace shale

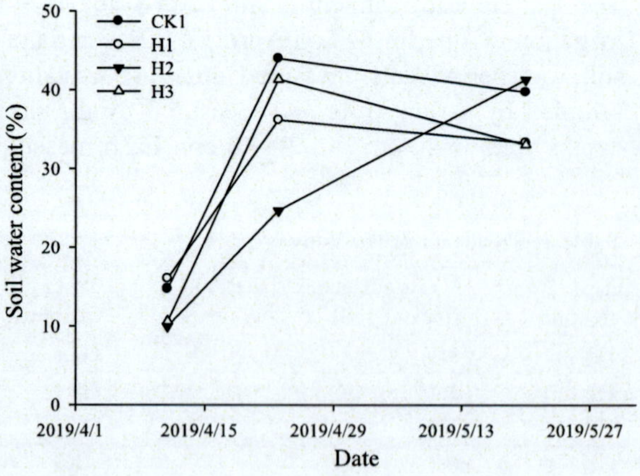

water of H1 and H2 was maintained at about 26 and 35%, respectively. In terms of 20–40 cm, the overall change trend is similar to 0–20 cm. Except for the H2 treatment, the water in the measurement on May 20 increased, and the other treatments showed a slight decline or a stable trend.

On April 23, the soil water content increased due to a previous irrigation, and CK1 and H3 decreased afterwards due to water evaporation and infiltration to the lower layer. The water content of H2 and H3 was twice later. Maintaining a relatively stable state can reflect that the increase in the content of argillaceous shale reduces the infiltration rate of the loessal soil and increases the water holding performance.

3.2 Influence of Argillaceous Shale Efflorescence on Ancient Soil Water Content

Figures 3 and 4 reflect the changes in soil water in mixed soils of paleosol and argillaceous shale. It can be seen that in the range of 0–20 cm soil layer, CK2 (Vs:Vp = 0:1) shows a trend of first increase and then stability, G1 (Vs:Vp = 1:1), G2 (Vs:Vp = 1:2) and G3 (Vs:Vp = 1:5) both show an increase first and then a small trend, in which CK2 is maintained at about 20.00%, G1 is reduced from 22.60 to 13.07%, a decrease of 9.53% G2 decreased from 11.76 to 9.35%, a decrease of 2.41%, and G3 decreased from 21.61 to 12.53%, a decrease of 9.08%. The 20–40 cm soil water content showed an increasing trend, and CK2, G1, G2, and G3 increased by 9.74, 18.20, −12.66, and −7.21%, respectively.

The above analysis shows that the response of soils to ancient soils, argillaceous shale mixed soils and yellow cotton, argillaceous shale mixed soils to water is different. The main reason is that the ancient soil itself is relatively viscous and the

Fig. 3 Variation of soil water content between 0–20 cm in mixed soil of ancient soil and argillaceous shale

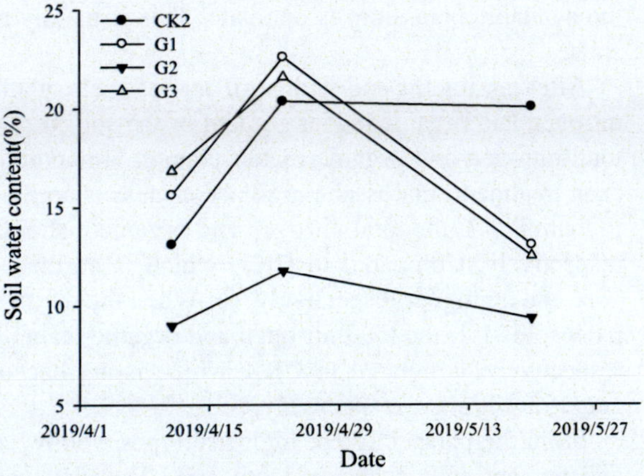

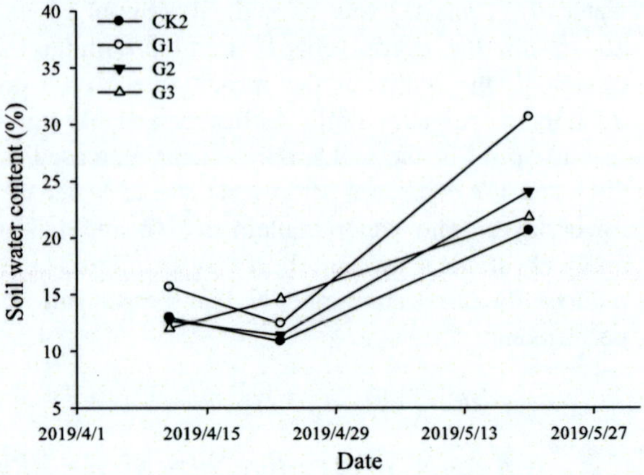

Fig. 4 Variation of soil water content of 20–40 cm in mixed soils of ancient soil and argillaceous shale

soil water infiltration process is relatively slow. The addition of argillaceous shale can improve the texture of the ancient soil to a certain extent, increase air permeability, and help water Infiltration.

3.3 *Physical and Chemical Properties of Soil Under Different Mixing Ratios of Mudstone*

It can be ground from Table 2. The soil conductivity, pH, and calcium carbonate did not change under different test treatments. The pH range of each treatment was between 8.4 and 8.7, and the calcium carbonate content was about 8%. Available phosphorus and available potassium have certain variability between different treatments. Overall, the effective phosphorus of the mixed soil of loess soil and argillaceous shale is higher than that of the ancient soil and argillaceous shale, while the available potassium is opposite. This is mainly related to the nature of the two soils.

After mixing the soil with argillaceous shale, it affects the penetration of the total nitrogen and organic matter content of the soil. It can be ground from Table 2. The total nitrogen and organic matter content is maximum when the ratio is 1:1 under each treatment, that is, the argillaceousness is argillaceous. When the ratio of shale to loam is 1:1, the total nitrogen and organic matter content are 0.77 and 5.03 g/kg, respectively. Compared to CK1, which is argillaceous shale:loam, 0:1, which is 93% and 26 higher respectively. %; When the mixing ratio of argillaceous shale and paleosol is 1:1, the total nitrogen and organic matter content are 0.86 and 3.45 g/kg respectively, compared to CK2, which is argillaceous shale:paleosol = 0:1 219% and 132% higher, respectively.

From the perspective of high-rise types, yellow cotton soil is powder loam soil, and ancient soil is clayey loam soil. The clay content is relatively high. After

Table 2 Physical and chemical properties of argillaceous shale and soil mixed soil

Treatments	Conductivity (ms/m)	pH	Available phosphorus (mg/kg)	Fast-acting potassium (mg/kg)	Total nitrogen (g/kg)	Organic matter (g/kg)
CK1	42.57	8.67	3.1	138	0.40	3.98
H1	34.44	8.68	2.7	108	0.77	5.03
H2	43.55	8.67	3.8	118	0.59	4.43
H3	39.48	8.47	3.2	146	0.50	3.53
CK2	24.96	8.56	2.4	183	0.27	1.49
G1	22.43	8.65	1.0	141	0.86	3.45
G2	27.37	8.53	2.9	152	0.62	1.99
G3	34.66	8.69	3.8	128	0.44	4.35

Treatments	Calcium carbonate content (%)	Clay (%)	Silt (%)	Sand (%)	Texture
CK1	8.244	14.26	82.63	3.11	Silt loam
H1	8.104	14.35	83.86	1.79	Silt loam
H2	7.718	13.85	84.94	1.21	Silt loam
H3	7.801	16.29	81.97	1.74	Silt loam
CK2	8.198	27.10	70.59	2.31	Silt clay loam
G1	7.666	23.31	72.65	4.04	Silt loam
G2	7.456	19.06	78.47	2.47	Silt loam
G3	7.924	18.30	77.16	4.54	Silt loam

mixing loessial soil with argillaceous shale, the content of sand grains was reduced from 3.11% to 1.21% −1.79%; while mixing of ancient soil with argillaceous shale reduced the clay content in the ancient soil to some extent The clay content was reduced to less than 27%, and the thickness of the soil changed from silty loam to silty loam.

The above results indicate that the addition of argillaceous shale is beneficial to increase the organic matter and total nitrogen content in the soil, increase the nutrients of cultivated land, and appropriately reduce the sand content of silty loam and the clay content of silty loam. The situation has potential application value.

4 Conclusions

This study is aimed at the phenomenon that a large amount of argillaceous shale is mixed during the land remediation process in Weibei, which leads to the change of soil water status. The field of argillaceous shale is studied by field sampling and indoor pot experiment of chemical compounds on water content in different types of soil. The main research conclusions are as follows.

The response of argillaceous shale efflorescence to the water content of loamy loess soil and partially viscous ancient soil is different. After mixing argillaceous shale with loamy loam, it reduces the water infiltration and soil water content to a certain extent, and it is also easier to maintain the stability of soil water and improve water holding performance; while mixing with clayey ancient soil Later, it is beneficial to increase water infiltration and soil water content. The main reason is that the ancient soil itself is relatively viscous, and the infiltration process of soil water is relatively slow. The addition of argillaceous shale efflorescence can improve the texture of the ancient soil to a certain extent, increase air permeability, and help. Infiltration of water.

The addition of argillaceous shale will help increase the organic matter and total nitrogen content in the soil to a certain extent, and increase the nutrients of the new cultivated land. The argillaceous shale can appropriately reduce the sand content of silty loam and the clay content of silty loam, and has potential application value for improving soil texture.

Acknowledgements This study was funded by the Internal research projects of Shaanxi Provincial Land Engineering Construction Group Co., Ltd., China (DJNY2018-19).

References

Yang, X. H., Jin, X. B., Guan, X., et al.: The spatial distribution of land consolidation projects in China from 2006 to 2012. Resources Science 35(08), 1535–1541 (2013).

Liu, Y. B., Li, X., Liu, Y. J., et al.: Construction method and application effect on tillage layer soil by sediment in land consolidation engineering. Transactions of the Chinese Society of Agricultural Engineering 31(09), 242–248 (2015).

Yang, F. B.: The post evaluation of land consolidation project in rural- based on the empirical land consolidation in Longchuanjiang basin in Nanchuan. Southwest University, Chongqing (2011).

Zhou, J., Zhang, F. R., Wang, X. L., et al.: Spatial- temporal change and analysis of land consolidation's newly increased cultivated land in China. Transactions of the Chinese Society of Agricultural Engineering 30(19), 282–289 (2014).

Zhou, B. B., Shao M. A., Wang, Q. J.: Effect of different rock fragments species on soil infiltration. Journal of Northwest A & F University(Natural Science Edition) 39(10), 141–148 (2011).

Mehuys, G. R., Stolzy, L. H., Letey, J., et al.: Effect of Stones on the Hydraulic Conductivity of Relatively Dry Desert Soil. Soil Science Society of America Journal 39(1), 37–42 (1975).

Xu, J. C., Shang, Y. Q.: Study on mechanism of disintegration and failure of shallow debris landslide under rainfall action. Journal of Natural Disasters (03), 117–124 (2008).

Wang, W. H., Wang, Q. J., Wang, S.: Experimental research on change characteristics of soil air permeability in stony soil medium. Transactions of the Chinese Society of Agricultural Engineering 28(04), 82–88 (2012).

Yang, Y. F., Wang, Q. J., Ceng, C., et al.: Experimental Research on Water Infiltration Characteristics of Stony-soil Medium. Journal of Soil and Water Conservation 23(05), 87–90 + 132 (2009).

Chen, Y. Q., Dong, Y. H. Progress of research and utilization of soil amendments[J]. Ecology and Environment 17(03): 1282–1289 (2008).
Hao X. Z., Zhou D. M., Xue Y., et al. Ryegrass growth in Cu mine tailings amended with natural montmorillonite and zeolite. Acta Pedologica Sinica 42(03):434–439 (2005).
Wang J. M., Yang P. L., Zhang J. G., et al. Salinity effect on sunflower at seedling stage during improving sodic soils reclaimed with by-product from flue gas desulphurization (BFGD). Transactions of the CSA E 21(09):40–44 (2005).

New Methodological Approach to Water Purification from Long-Lived Radionuclides and Heavy Metals Under Emergency Conditions

Lydia Bondareva and Nataliia Fedorova

Abstract The presented research results show a possibility of using a synthesized sorption material impregnated with iron (III) oxyhydroxide and manganese (IV) dioxide fixed on a polyurethane foam sheet and forming a nanostructure in the whole volume of this sheet. The above material can be used for the purposes of water purification from radionuclides and heavy metals to reach the level of potable water. The research was carried out using: (1) model systems obtained by mixing metal salt solutions and radionuclides: (Pb^{2+}—5 mg/l, Hg^{2+}—1 mg/l, Cd^{2+}—10 mg/l, Cu^{2+}—50 mg/l, ^{238}U—10 mg/l, ^{137}Cs—100 Bq/l, ^{242}Pu—5 Bq/l) and (2) water samples from surface water reservoirs. When compared, the two methods of purifying model water samples (sorption and co-precipitation) show consistent results. The residual content of the introduced metals and radionuclides varied within the error of the techniques of sample preparation and detection. However, the application of the developed sorbent showed considerable advantages due to its rapidity and ability to simplify the water purification process which is especially valuable for the application under field conditions. The suggested scheme of water purification can be used for effective water treatment including the conditions of emergencies.

Keywords Sorption · Radionuclides · Heavy metals · Water

1 Introduction

The impact of water quality on the health of population is an extremely urgent issue. In industrially developed countries the requirements to the criteria of potable water quality are higher, which reflects the level of technologies and possibilities available to provide proper purification of potable water under the conditions of the

L. Bondareva (✉) · N. Fedorova
Federal Scientific Center named after F. Erisman, Mytischi, Moscow Region 141014, Russia
e-mail: lydiabondareva@gmail.com

R. G. Gutierrez (ed.), *Wastewater Technologies and Environmental Treatment*, Springer Proceedings in Earth and Environmental Sciences,
https://doi.org/10.1007/978-3-030-61989-3_2

increasing natural and man-made pollution of water sources (Guidelines on sanitation 1470; Stocks et al. 2014; Penakalapati et al. 2017).

More than 400 kinds of substances have been established to cause water pollution. In the case of exceeding the permissible level as concerns at least one of the three criteria of harmfulness: sanitary-toxicological, general sanitary or organoleptic one, water is considered to be polluted. One can distinguish chemical, biological and physical pollutants (Fuller and Eisenberg 2016; Borzunova and Kuzmin 2007). Among the chemical pollutants the most widely spread ones include oil and oil products, synthetic surfactants, pesticides, heavy metals, dioxins etc. From the viewpoint of water pollution biological pollutants are also very dangerous: viruses and other pathogenic microorganisms; as well as physical factors, such as radioactive substances, heat etc. Pollution of aquatic ecosystems is of great danger for all living organisms of great danger for all living organisms, especially dangerous is it for man (Novikov 2002; Sadovnikova et al. 2008). Unfavorable for human health consequences of using polluted water appear either after direct contact (swimming, washing, fishing etc.), as well as after drinking or as a result of biological accumulation (Guidelines on sanitation 1470; Borzunova and Kuzmin 2007; Korolev et al. 2014).

This is of great importance under the conditions of emergency. All possible types of emergencies, most of them being unpredictable, result in irreversible consequences, with the most important one being danger to health and life of the population in a disaster area (Muchkinova and Turdumatov 2013; State Standard RF 22.0.05-94; Akimov 2007; Kryuchek and Latchuk 2005). Emergencies include man-made accidents at extractive and processing enterprises as well as terrorist attacks directed towards putting out of operation the main water supply system (Muchkinova and Turdumatov 2013; State Standard RF 22.0.05-94; Kryuchek and Latchuk 2005). Besides, there can be natural disasters which can prevent the population access to potable water (Akimov 2007; Kryuchek and Latchuk 2005; Mashkova 2007).

In the direct contact with bacteriologically polluted water, as well as in the case of living or staying near a water body various parasite can penetrate through the skin and cause serious diseases, especially characteristic of tropical and subtropical regions. Under the modern conditions there is an increasing danger of such epidemic diseases as cholera, typhoid, dysentery etc.

The development of nuclear power industry is directly connected with the most important condition of providing its safety, which was evidenced by one of the biggest radiation accidents at a nuclear power plant Fukushima-1 (Japan) in 2011. Experts in the field of nuclear safety from the leading world countries declare safety of nuclear power industry to be above and beyond national borders and suggest that the requirements to operating nuclear power facilities should be considerably tightened. As a result of unforeseen emergencies at the enterprises of nuclear fuel cycle (Fukushima-1, Chernobyl nuclear power plant) the adjacent sources of water supply become unsuitable for the direct purpose of providing potable water for the population. This, in its turn, results in social strain among the population and possible imbalance in the society.

The literature analysis shows that adsorbents based on zeolites (Tager 1968) and silicon oxide (Kargin 1977), as well as those based on synthetic polymer materials (acrylic or polypropylene fibers, cellulose etc.) (Lisichkina 1986; Golik et al. 2011; Myasoedov 2005) are widely used for the purification of natural aquatic media from radionuclides. For commercial purposes use is made of columns and discs filled with an organic phosphorus adsorbent UTEVA (Eichrom, USA) to purify water bodies by concentrating uranium and thorium (Milyutin 2008). The efficiency of extraction in all the cases does not exceed 65%. In order to achieve this efficiency, pre-treatment of the water samples is necessary as well as a long procedure of interaction and, most often, the extraction process needs the laboratory conditions.

Besides, some other methods of concentration are used followed by the purification of water samples; these methods are based on the precipitation or co-precipitation of radionuclides on finely dispersed or colloidal particles formed in a water sample or introduced into samples as a ready mixture (Batuk et al. 2011; Macrocyclic compounds in analytical chemistry 1993).

Another technique of purifying industrial discharges from metals is known which is based on a multistage water filtration (Yakovlev et al. 1985). The disadvantages of this liquid filtration technique include: limited rate of the liquid flow on a particular stage, insufficiently firm adherence of the filtering material to the walls of the filter body, which can result in the penetration of the primary solution into the area with the purified liquid without passing through the filtering material.

The aim of the present study is to study the process of water purification from radionuclides and heavy metals using nanostructured particles of a sorbent based on a mixture of iron oxyhydroxide and manganese oxide impregnated into a polyurethane foam sheet.

2 Objects and Methods

2.1 Preparation of Sorption Material

In the study use was made of polyurethane foam sheets of the type ST22/40, with the size of 10 × 500 × 1000 mm, produced by Europlast (Russia). Iron salt $FeCl_3 \cdot 6H_2O$, pure, (State Standard 4147-74), Russia; $KMnO_4$ (CAS: 7722-64-7), reagent grade, Component-reaktiv Ltd., Russia; $MnCl_2 \cdot 2H_2O$, reagent grade (State Standard 612-75, 1261001), Russia were used as reagents. The sorbent is a synthetic material preliminarily impregnated with a mixture of inorganic substances, Fe(III) oxyhydroxide and manganese oxide MnO_2.

The prepared substrate, a polyurethane foam (PUF) sheet with the thickness of 1 cm and size 100 × 100 cm, with the help of tweezers was carefully and slowly placed into a suitable container, for example, evaporation dish with the volume of 3 l, preliminarily filled with a 3 M HCl solution in the volume of 1.5 l where it was kept for 30 min. Then, this sheet was pressed out using special cylinders and rinsed with a large amount of running water to remove dissolved and mechanical contaminants. Afterwards, the washed and pressed PUF sheet was again placed into the evaporation dish with the fresh acid solution where it was completely impregnated with HCl by regular stirring for 60 min. Further, excessive acid was removed from the PUF sheet and the sheet was placed into another evaporation dish filled with a solution of iron salt (II) (the required portion of iron salt (III) was dissolved in a 3 M HCl solution, with the final iron concentration being 15 μg/l). The polyurethane foam sheet impregnated with the iron salt acquired the light-yellow color. Further, the PUF sheet was slightly pressed out to remove the excess of the reagent and placed into ammonium water (distilled water with ammonium added to reach pH $\sim$ 9) until complete wetting with continuous stirring with a glass stick. The oxidation of iron (II) to iron (III) occurred with the formation of iron oxyhydroxide and the whole volume of the PUF sheet acquiring the light brown color.

The colored polyurethane foam sheet was placed into a bath filled with potassium permanganate solution, with the concentration being 1 g/l. The PUF sheet was completely immersed into the solution and with stirring it evenly acquired the violet color. Then, almost without pressing out the sheet was placed into an evaporation dish containing 3 l of the solution of $MnCl_2$ salt with the concentration of 40 g/l, where the final stage of the sorbent formation occurred, which was indicated by the change of the color into dark brown due to the reaction of reduction of the permanganate ion into manganese dioxide. In the final solution the PUF sheet was very carefully impregnated until the violet color completely disappeared. The duration of the process was 30–45 min.

The ready sheet was pressed out, carefully rinsed with water to remove free sediment particles, and then, it was placed into a drier for 48 h at 30 °C.

After drying, discs suitable for a specific filtering device were cut from the PUF sheet using a press form (with the device being available and the discs being no bigger than 10 cm in diameter) to be later packed for transportation or storage in appropriate containers. In the present research, the filtering device developed by L. Bondareva and described in Patent RU (2013) was used.

The storage time of the obtained material is 5 years. The sorption material is stable in the range of ambient temperatures from −40 °C to +80 °C, and it is easy to use and to dispose of. It has a low cost and it is not deformed during its preparation and filtering. Before use, the material must be moistened with water and allowed to stand for 10 min. Before filtering, pH in the water sample must be adjusted to 7–8.

The advantage of the obtained sorbent is its finished form which can be placed in a sorption container without additional preparation. In this case, the shape can be set arbitrarily, depending on the filtering device used.

When using the obtained structured sorption material for water purification from radionuclides, two radioactive isotopes ^{241}Am (12 Bq/sample) and ^{236}Pu (8 Bq/sample) were introduced into settled tap water; based on their content the capacity of the sorption material was estimated in the final filtrate.

The chemical yield for ^{241}Am was 95 $\pm$ 2%, and for ^{236}Pu it was equal to 76 $\pm$ 5%.

2.2 Co-precipitation Technique

To confirm that the applied water purification technique was efficient, precipitation of Fe oxyhydroxide was first performed for comparison and then, after its separation, manganese dioxide precipitated directly in the river water sample.

For this, 20 ml of HNO_3 (concentrated), a solution of $^{241}Am(NO_3)_3$ (12 Bq/sample) and $^{236}Pu(NO_3)_4$ (8 Bq/sample), 5 ml of a solution of Fe(III) ($FeCl_3$) (C_{Fe}^{3+} = 5 Bq/ml) were added to 20 ml of the water sample in a transparent plastic bottle. The sample thus obtained was mixed and allowed to stay overnight to establish equilibrium. Then, a solution of NH_4OH (25%) was introduced to the sample up to pH = 8–9, and 1 ml of $KMnO_4$ solution (60 g/l) was added and immediately after stirring, other 0.2 ml of $MnCl_2$ solution (40 g/l) was introduced. The mixture was again thoroughly stirred and allowed to stand for 24 h for precipitation. Then, a larger part of the liquid above the precipitate was removed by decantation. The remaining liquid was separated from the precipitate through a paper filter with the pore diameter of 3.5 μm. To detect ^{241}Am, the precipitate on the filter was placed in a special polyethylene vessel—«denta», and analyzed using a gamma spectrometer as described in Bondareva et al. (2008). The precipitate obtained was used to detect ^{236}Pu by the radiochemical method of extracting isotopes (Methodical Recommendation 2.6.1.0064-12; Talvitie 1971); after electrolytic deposition onto a substrate, the radionuclide was detected by α-spectrometry.

The chemical yield for ^{241}Am was 92 ± 2%, and for ^{236}Pu it was equal to 71 ± 5%.

2.3 Methods of Analysis

To estimate the concentration of metals and radionuclides in the initial and final solutions as well as on the sorption material we used an Agilent 7500a Quadrupole mass spectrometer (Agilent Technologies, USA) and gamma-spectrometry with the help of a Canberra spectrometer (USA) with an ultrapure germanium detector (HPGe). Plutonium isotopes were detected by α-spectrometry using an α-spectrometer 7184 (Eurisys Mesures, France).

Electron microscopy studies were performed using a scanning electron microscope TM-1000 (Hitachi, Japan) with an energy-dispersive X-ray analyzer SwiftED (Oxford Instrument Analytical, England), operating in back-scattered electron mode at an accelerating voltage of 15 kW in low-vacuum. Moreover, electron microscopy studies were performed using a scanning electron microscope JEOL JSM-7001F with an energy-dispersive X-ray spectral analyzer Oxford Instrument INCA PentaFETx3. For this, filters were coated with a 15 nm layer of gold. An Au layer was made in high-vacuum using JEOL JEE 420.

3 Results and Discussion

To reveal the efficiency of the water purification from radionuclides and metals a series of model experiments were performed simulating the water contamination with a number of toxic metals and radionuclides (Pb^{2+}—5 mg/l, Hg^{2+}—1 mg/l, Cd^{2+}—10 mg/l, Cu^{2+}—50 mg/l, ^{238}U—10 mg/l, ^{137}Cs—100 Bq/l, ^{242}Pu—5 Bq/l). For this purpose, salts of the studied metals and radionuclides were added to a container with drinking water (V = 19 l) after settling to remove dissolved chlorine. The studies were carried out using both the preliminarily obtained sorption material and direct co-precipitation of the introduced metals and radionuclides from the water sample (Table 1) according to the co-precipitation technique described in the experimental part as well as in Bondareva et al. (2008).

As can be seen from the presented results the difference between the two methods is negligible, being within the measurement error.

However, using the ready-made sorbent material to purify the water revealed significant advantages, the most important being: (1) reduction in the duration of the water treatment (about 3–5 h), as compared with the co-precipitation method (up to 4 days), (2) ease of use, especially in field conditions. In addition, the technology for producing the sorbent material based on polyurethane foam allows one to use it for any filtering device available in field and stationary laboratories.

Taking into account that the introduced concentrations of the studied metals and radionuclides significantly exceeded the maximum allowable concentration for the content of chemical substances in water which was established for water bodies for drinking, cultural and domestic use (State regulations 1315), we suggested the following. The proposed method for purifying such concentrations will be more efficient for drinking water and will allow obtaining values which will satisfy the sanitary and hygienic standards for drinking water.

To confirm this, three water sources were chosen which differed in their qualities: (1) water of the Yenisei River (ship channel) upstream of the city of Krasnoyarsk; (2) water of the reservoir on the Bugach river located on the north-eastern outskirts of the city of Krasnoyarsk; (3) water of the Yenisei River, taken in the impact zone of the Mining and Chemical Combine of the Rosatom State Corporation, 75 km downstream of Krasnoyarsk (Bondareva et al. 2008). The initial content of some heavy metals and radionuclides in the studied water samples is shown in Table 2.

As can be seen from the data presented, all the water samples studied were found to contain significant concentrations of heavy metals. Moreover, water sample № 3 contained a wide range of man-made radionuclides.

A source of the release of radionuclides into the water was likely to be the operation of the Mining and Chemical Combine of the Rosatom State Corporation whose main activity was the production of weapons-grade plutonium and whose nuclear reactors operated for a long time (Bondareva et al. 2008).

Using both methods to purify the samples, it was found that the content of radionuclides in the water was smaller than the limit of detectable activity, except

Table 1 Results of estimating residual amounts of the introduced metals and radionuclides in the water of the model systems after the purification using the methods of sorption (I) and co-precipitation (II) in mg/l (%), n = 5, p = 0.95

	Pb^{2+}		Hg^{2+}		Cd^{2+}		Cu^{2+}	
	I	II	I	II	I	II	I	II
mg/l	0.05 ± 0.01	0.07 ± 0.02	0.010 ± 0.005	0.02 ± 0.01	0.14 ± 0.04	0.20 ± 0.08	0.9 ± 0.3	1.4 ± 0.5
%	1.0 ± 0.2	1.4 ± 0.4	1.0 ± 0.5	2.0 ± 1.0	1.40 ± 0.4	2.0 ± 0.8	1.8 ± 0.6	2.8 ± 1.0
	^{238}U			^{137}Cs			^{242}Pu	
	I	II		I	II		I	II
mg/l (Bq/l)	1.0 ± 0.4	1.5 ± 0.6		(5.3 ± 1.2)	(6.8 ± 2.1)		(1.25 ± 0.45)	(1.45 ± 0.56)
%	10 ± 4	15 ± 6		(5.3 ± 1.2)	(6.8 ± 2.1)		(25 ± 9)	(29 ± 11)

Table 2 Chemical composition of the initial water samples taken for the research

Samples	Chemical composition							
	Fe (mg/l)	Sr (mg/l)	Mn (mg/l)	Ti (mg/l)	Zn (mg/l)	Cu (mg/l)	Ni (mg/l)	V (mg/l)
1[a]	35 ± 13	2.7 ± 1.0	0.06 ± 0.01	0.002 ± 0.001	0.034 ± 0.011	0.005 ± 0.001	0.002 ± 0.001	0.009 ± 0.002
2[b]	68 ± 24	14 ± 3	0.13 ± 0.05	0.004 ± 0.001	0.086 ± 0.021	0.017 ± 0.007	0.021 ± 0.009	0.019 ± 0.008
3[c]	125 ± 21	38 ± 7	1.17 ± 0.32	0.011 ± 0.004	1.41 ± 0.07	0.093 ± 0.007	0.069 ± 0.011	0.053 ± 0.011
	Cr, mg/l	Pb, mg/l	Co, mg/l	Cd, mg/l	Hg, mg/l	^{24}Na, Bq/l	^{46}Sc, Bq/l	^{51}Cr, Bq/l
1[a]	0.006 ± 0.002	0.001 ± 0.0005	0.027 ± 0.003	0.006 ± 0.002	0.001 ± 0.0005	<MDA[d]	<MDA	<MDA
2[b]	0.011 ± 0.004	0.009 ± 0.002	0.034 ± 0.005	0.012 ± 0.004	0.009 ± 0.002	<MDA	<MDA	<MDA
3[c]	0.036 ± 0.012	0.051 ± 0.011	0.097 ± 0.012	0.047 ± 0.011	0.013 ± 0.008	250 ± 14	0.24 ± 0.01	6.7 ± 0.2
	^{54}Mn, Bq/l	^{59}Fe, Bq/l	^{60}Co, Bq/l	^{65}Zn, Bq/l	^{76}As, Bq/l	^{85}Sr, Bq/l	^{99}Mo, Bq/l	^{103}Ru, Bq/l
1[a]	<MDA	<MDA	<MDA	<MDA	<MDA	<MDA	<MDA	<MDA
2[b]	<MDA	<MDA	<MDA	<MDA	<MDA	<MDA	<MDA	<MDA
3[c]	0.09 ± 0.02	0.11 ± 0.01	0.2 ± 0.1	0.11 ± 0.01	8.5 ± 0.6	0.02 ± 0.01	0.09 ± 0.01	0.027 ± 0.004
	^{106}Ru, Bq/l	^{124}Sb, Bq/l	^{137}Cs, Bq/l	^{141}Ce, Bq/l	^{144}Ce, Bq/l	^{239}Np, Bq/l	U_{total}, mg/l	^{238,239}Pu, Bq/l
1[a]	<MDA	<MDA	36 ± 4	<MDA	<MDA	<MDA	1.5 ± 0.6	<MDA
2[b]	<MDA	<MDA	61 ± 7	<MDA	<MDA	<MDA	4.0 ± 0.8	<MDA
3[c]	0.08 ± 0.03	0.024 ± 0.003	175 ± 9	0.05 ± 0.01	0.13 ± 0.02	17.1 ± 0.3	13 ± 3	3 ± 1

[a] site 1 (water of the Yenisei River (ship channel) upstream of the city of Krasnoyarsk)
[b] site 2 (water of the reservoir on the Bugach river located on the north-eastern outskirts of Krasnoyarsk)
[c] site 3 (water of the Yenisei River taken downstream of the discharges of the Mining and Chemical Combine of the Rosatom State Corporation, 75 km downstream of Krasnoyarsk)
[d] <MDA (minimum detected activity)

for ^{24}Na and ^{137}Cs, which are not adsorbed nor co-precipitate due to the peculiarities of their physicochemical properties, mainly due to the absence of possible insoluble compounds of cesium and sodium. The content of most metals in the samples under study after purification was within the limit of detection, i.e. lower than 1–10 μg/l.

The surface morphology of the obtained sorption materials was studied by electron microscopy (SEM). Figures 1, 2 and 3 show SEM-images of the initial sample of polyurethane foam (Fig. 1), after its preparation for sorption (Fig. 2) and after the sorption (Fig. 3a, b). The fragment presented in Fig. 3b is worth paying special attention to.

As is seen, the surface of the particle is covered with small nanoscale fragments (d = 10–100 nm), which are likely to be the substances precipitated from the water

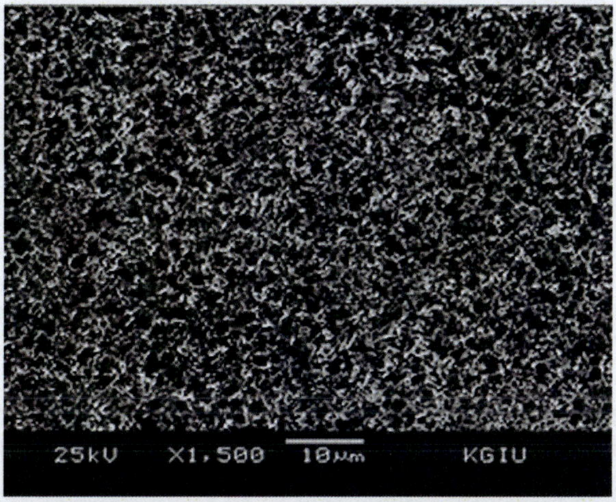

Fig. 1 SEM-image of the initial sample of polyurethane foam (PUF), ×1500

Fig. 2 SEM-image of the impregnated sample of polyurethane foam, ×5000

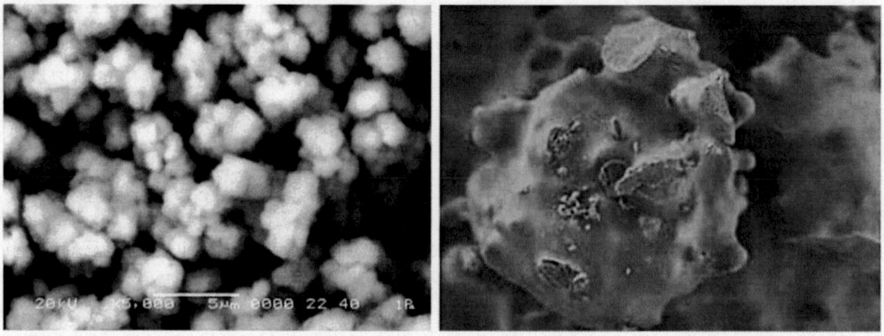

Fig. 3 SEM-image of the impregnated PUF sample after sorption: **a** ×5000, **b** ×10000

samples consisting of a mixture of iron oxyhydroxide and manganese oxide. Moreover, these particles are located in the entire layer of polyurethane foam.

4 Conclusion

To sum up, the proposed synthesized nanostructured sorption material impregnated with a mixture of iron oxyhydroxide and manganese oxide (IV) can be used to effectively prepare water samples for different needs of the population, including drinking water as well as that in emergency situations. The main advantages of the proposed sorption material are ease of use, efficiency of purification, preservation of the desired properties during long-term storage, ability to give it any shape of the existing filtering devices.

Acknowledgements The study was partially supported by the specialized program "Basic research" of the Federal Service for Surveillance of Consumer Rights Protection and Human Well-being.

References

Guidelines on sanitation and public health. ISBN 978-92-4-451470-2 (In Russ).

Stocks ME, Ogden S, Haddad D, Addiss DG, McGuire C, et al. Effect of Water, Sanitation, and Hygiene on the Prevention of Trachoma. A Systematic Review and Meta-Analysis (5), 254–261 (2014).

Penakalapati G, Swarthout J, Delahoy MJ, McAliley L, Wodnik B, Levy K et al. Exposure to animal feces and human health: a systematic review and proposed research priorities. Environ Sci Technol. 51(20), 11537–11552 (2017).

Fuller JA, Eisenberg JN. (2016). Herd protection from drinking water, sanitation and hygiene interventions. Am J Trop Med Hyg. 95 (5), 1201–1210 (2016).

Borzunova EA, Kuzmin SV, Akramov RL Assessing the impact of drinking water quality on public health. Hygiene and sanitation, 3, 32–34. (2007).

Novikov Yu.V. Ecology of the environment. M.: Science (2002) (In Russ).

Sadovnikova L.K., Orlov D.S., Pozanovskaya I.N. Ecology and environmental protection during chemical pollution. M.: Higher school. (2008) (In Russ).

Korolev A.A., Bogdanov M.V., Korolev Al.A. Medical Ecology/ Ed. prof. A.A. Koroleva. - 3rd ed., Revised. and add. - M.: Academy (2014) (In Russ).

Muchkinova L.I., Turdumatov B.M. (2013) Manual on technogenic emergencies: textbook. student manual [Electronic resource]/. — Elista: Kalmyk State University, (2013) https://rucont.ru/efd/299096 (In Russ).

State Standard RF 22.0.05-94. Safety in emergency situations. Man-made emergencies. Terms and Definitions. http://docs.cntd.ru/document/1200133493 (In Russ).

Akimov V.A. (2007) Safety in natural and man-made emergencies: Textbook. allowance/ -2nd ed., revised. M.: Higher school. (2007) (In Russ).

Kryuchek N.A., Latchuk V.N. (2005) Safety and protection of the population in emergency situations: a study method. Allowance for classes with the population/ Under the total. ed. G. N. Kirillova. M.: NTs ENAS, (2005) (In Russ).

Mashkova S. A. (2007) Wastewater treatment using natural sorbents and their chemically modified analogues (PhD dissertation). Moscow. (2007) (In Russ).

Tager A.A. (1968) Physical chemistry of polymers. M.: Chemistry. (1968) (In Russ).

Kargin V.A. Encyclopedia of Polymers. T 3. - M.: Soviet Encyclopedia, (1977) (In Russ).

Lisichkina G.V. (1986). Modified silicas in sorption, catalysis and chromatography. M.: Chemistry. (1986). (In Russ).

Golik V.M., Golik S.V., Ivanov S.L. (2011) Development of an inductively coupled plasma mass spectrometric method for the analysis of uranium hexafluoride with chromatographic separation of uranium and impurities on a column with UTEVA resin// Analytics and Control. 15 (2), 174–181. (2011) (In Russ).

Myasoedov B.F. (2005) Methods of separation and concentration in solving urgent problems in radiochemistry. Russian Chemical Journal. (2). 64–67. (2005) (In Russ).

Milyutin V.V. Physico-chemical methods for the extraction of radionuclides from liquid radioactive waste of low and medium levels of activity (Doctor dissertation). Moscow. (2008) (In Russ).

Batuk D.N., Shiryaev A.A., Kalmykov S.N., Batuk O.N., Romanchuk A.YU, Shirshin E.A., Zubavichus Y.V. (2011) Sorption and speciation of uranium on silica colloids. In book: actinide nanoparticle research. GERMANY: SPRINGER HEIDELBERG. R., 315–322. (2011).

Macrocyclic compounds in analytical chemistry (1993)/ Ed. Yu.A. Zolotova, N.M. Kuzmina. M.: Science. (1993) (In Russ).

Nikitina T.V., Sabgaida H.A., Olshanskaya L.N. Wastewater treatment from heavy metal ions and oil products by composite filters. On Sat Promising polymer composite materials. Alternative technologies. Recycling. Application. Ecology. Reports of the 5th International Conference "Composite - 2010". Saratov. S. 446–448. (2010) (In Russ).

Yakovlev S.V., Karelin J.A., Laskov Yu.M. (1985). Purification of industrial wastewater/ C. In *Revised. and add.* M.: Stroyizdat, 1985. (In Russ).

Patent RU 2484021 C1, 06/10/2013. (In Russ).

Bondareva L.G., Bolsunovsky A.Ya., Trapeznikov A.V., Degermendzhi A.G. Using a new method for concentrating transuranic elements in water samples of the River Yenisei. Reports of Chemistry. 423 (4). 479–482. (2008).

Methodical Recommendation 2.6.1.0064-12. Ionizing radiation, radiation safety. Radiation monitoring of drinking water by radiochemical analysis methods. Methodical recommendations: Radiochemical isolation of radionuclides and preparation of counting samples to determine their specific activity in water samples. http://www.consultant.ru/document/cons_doc_LAW_131734/4a30ce934f52a128297ebd474f9199af3aa24a1e/ (In Russ).

Talvitie N.A. Radiochemical determination of plutonium in environmental and biological samples by ion exchange. Analytical Chemistry, 43 (13), 1827–1830. (1971).

State regulations 2.1.5.1315-03. The extent allowed concentrations of chemicals in water of water facilities for economic drinking and cultural-water use. (In Russ).

Analysis of the Effects of Flood Control Constructions for Harbin Under Different Typical Historical Flood Conditions

Gongxun Guan, Changchun Ding, Ming Gao, and Lei Zhang

Abstract The effectiveness of two reservoirs constructed for flood control for Harbin were analyzed using historical flood data. These analyses demonstrated that the effectiveness of Baishan and Fengman reservoirs vary with the type of flood. They significantly reduced flood peaks when the flood mainly came from the basin above Fengman reservoir, while they had little effect when the flood mainly came from the upper and middle reaches of Nen River. The multi-reservoirs in the Nen river basin and Second Songhuajiang river basin contributed minimally to flood control for Harbin. Pangtoupao and Yueliangpao retention basins played important roles in flood control for Harbin. However, analysis of recent rainfall data of the area indicates that rainfall intensity in the flood season has significantly increased compared with that in the 2000s. Consequently these flood control projects cannot meet future flood control requirements of Harbin.

Keywords Harbin · Flood control · Typical flood · Precipitation

1 Introduction

Harbin is located in the southeast of Songnen Plain, on the banks of the middle reaches of Songhua River. As the capital of Heilongjiang Province, Harbin is not only an important industrial base in China, but also the largest transportation hub in Northeast China, whose road transportation network extends in all directions (Hu and Pan 2010). Songhua River is also connected to the Sea of Japan through the Heilongjiang River. Due to its important role in transportation, Harbin has devel-

G. Guan · C. Ding · M. Gao
Heilongjiang Water Conservancy and Hydropower Survey and Design Institute, 150080 Harbin, China

L. Zhang (✉)
State Key Laboratory of Simulation and Regulation of Water Cycle in River Basin, China Institute of Water Resources and Hydropower Research, 100038 Beijing, China
e-mail: lei207b@163.com

R. G. Gutierrez (ed.), *Wastewater Technologies and Environmental Treatment*, Springer Proceedings in Earth and Environmental Sciences,
https://doi.org/10.1007/978-3-030-61989-3_3

oped rapidly in recent years. Consequently, its need for flood control has become increasingly important. However, the current flood control system in Harbin is not yet fully developed and problems remain that need further research and discussion (Wang et al. 2016).

Previous studies have described and evaluated the projects, systems, and schemes for flood control in Harbin (Wu 2004; Ma 2005; Zhao 2006; Meng and Zang 2011; Zhang 2016; Chen 2004), and have summarized and considered existing problems (Wang et al. 2016; Han 2007; Yuan and Jin 1993). However, there is no quantitative analysis of the role played by existing flood control structures for different types of typical historical floods. In the context of to global climate change, extreme precipitation events may occur frequently all over the world, including within the catchment of Harbin (Jiang et al. 2007; Yang et al. 2008; Zhai et al. 2007). The frequency and extent of major floods in Harbin in the future remains unknown. Consequently, based on previous studies and historical flood data of Songhua River Basin, this paper quantitatively analyzes the effects of the current flood control structures in mitigating floods of Harbin, aiming to achieve a clearer understanding of the flood control capacity of Harbin and its catchment. Furthermore, rainfall data of Songhua River Basin in recent years were analyzed to (i) establish how rain intensity upstream of Harbin changes seasonally and inter-annually; and (ii) determine the rainfall characteristics of Harbin in different periods, thereby providing a reliable and reasonable scientific basis for the establishment of flood control structures in the future.

2 Typical Floods in Recent History

The flood season of the Songhua River Basin is from June to September and major floods generally occur in July and August. Floods in the region are frequent, often occurring once every two years. After the founding of the People's Republic of China, floods in 1953, 1956, 1957, 1960, 1969, 1998, 2010, and 2013 in the Songhua River Basin were considered to be representative. The measured peak flow of each flood is shown in Table 1. The floods in 1969, 1998, and 2013 were selected as typical floods in the Nen River Basin; the floods in 1953, 1956, 1960, 1991, 1995, 2010, and 2013 were selected as typical floods in the Second Songhua River Basin, and the floods in 1956, 1957, 1960, 1969, and 1998 were selected as typical floods in the main stream of Songhua River.

3 Major Flood Control Projects

After many years of construction, a basic flood control system was installed for Harbin, consisting of large reservoirs, including Baishan Reservoir and Fengman Reservoir, retention basins, including Pangtoupao Retention Basin and

Table 1 Maximum measured peak flows of typical floods in different years

Year	Measured flood peak flow (at Harbin station) (m^3/s)	Rainfall area
1953	9,530	Second Songhua River
1956	11,700	Nen River, Second Songhua River, Lalin River
1957	12,200	Nen River, Second Songhua River
1960	9,140	Main stream of Songhua River
1969	8,500	Nen River
1991	10,700	Second Songhua River
1995	6,120	Second Songhua River
1998	16,600	Nen River
2010	5,370	Second Songhua River
2013	10,400	Upper reaches of Nen River, upper reaches of Second Songhua River, some tributaries of the main stream of Songhua River

Yueliangpao Retention Basin, as well as embankment dams (e.g. Nierji dam). The locations and characteristic indices of representative reservoirs and flood retention basins upstream of Harbin are shown in Table 2, respectively.

4 Analysis of Effects Due to Major Flood Control Projects in Different Typical Floods

The effects of the aforementioned major flood control structures on typical floods in different years vary greatly as these structures are limited by their geographical locations.

Baishan Reservoir and Fengman Reservoir played an important role only when major floods occurred in the second Songhua River or the entire river basin (Nen River, Second Songhua River, and Songhua River). When major floods occurred in the Nen River Basin or the main stream of Songhua River, the Baishan Reservoir and Fengman Reservoir not only have little effect, but also pose further flood control challenges downstream (that is, Harbin) if they are adjusted improperly. The maximum inflow and maximum discharge volumes of Fengman Reservoir and the peak flow at the downstream Fuyu station, as well as the measured peak flow at the Harbin station, are summarized for typical historical floods in the second Songhua River Basin in 1953, 1956, 1960, 1991, 1995, 2010, and 2013 in Table 3 and Fig. 1. The results demonstrate that:

1. Except for the flood in 1956, the average peak reduction ratio of Fengman Reservoir with other flood conditions was 33.19%, demonstrating a significant reduction in peak flows. This was attributed to Baishan Reservoir and Fengman Reservoir, which reduced the peak flow at Harbin station to 9530, 6120, and

Table 2 Characteristics of reservoirs and major flood retention Basins (Evaluation Report on the Status of Harbin Urban Flood Control System 2019)

Characteristic index	Unit	Baishan	Fengman	Nierji	Pangtoupao	Yueliangpao
Watershed area above dam	km^2	19,000	42,500	66,382		
Average annual runoff at dam	10^8 m^3	74.03	135.3	104.7		
Average annual flow at dam	m^3/s	235	429	332		
Total storage capacity	10^8 m^3	59.07	103.77	8.11	45.647	22
Dam/dike crest elevation	m	423.5	269.5		133.17	136.57
Check flood level	m	420	268.5	219.9		
Design flood level	m	418.3	268.2	218.15		
Normal water level	m	413	263.5	216	131.67	134.57
Limit water level (flood season)	m	413	260.5	213.37		
Backwater level	m	380	242	195		
Installed capacity/unit	MW/unit	1,500/5	1,480/8	250/4		

Table 3 Effect of Fengman reservoir for different typical floods

Year	Maximum inflow (m^3/s)	Maximum discharge or peak flow at Fuyu station (m^3/s)	Peak reduction ratio (%)	Peak flow at Harbin station (m^3/s)	Remark
1953	17,700	6,550	37.01	9,530	Maximum discharge
1956	4,660	4,230	90.77	11,700	Peak flow at Fuyu station
1960	10,400	4,200	40.38	9,140	Peak flow at Fuyu station
1991	11,500	3,990	34.70	10,700	Peak flow at Fuyu station
1995	13,500	4,620	34.22	6,120	Maximum discharge
2010	19,700	5,580	28.32	5,370	Peak flow at Fuyu station
2013	10,700	2,620	24.49	10,200	Peak flow at Fuyu station

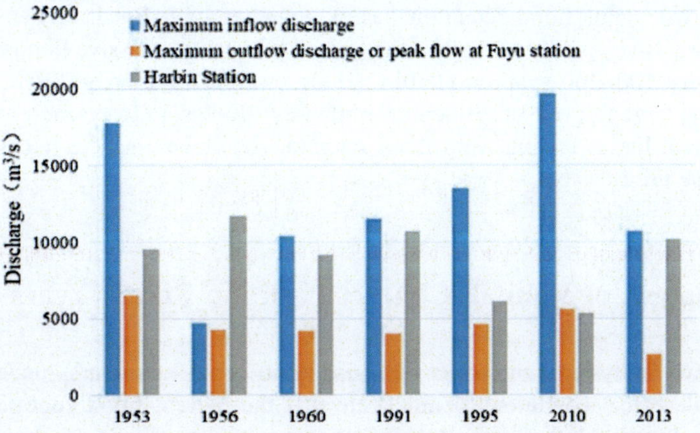

Fig. 1 A comparison of peak flows from the Fengman reservoir under flood conditions in different years

5370 m^3/s in 1953, 1995, and 2010, respectively, thereby greatly alleviating the flood control challenge in Harbin.

2. The reason for the peak reduction ratio of Fengman Reservoir being as high as 90.77% in 1956 was the lack of experience in flood control of this reservoir at that time. The limit water level was set as high as 263.50 m, which led to the large discharge of the reservoir in 1956.
3. Although Fengman Reservoir played a significant role in other years, due to the simultaneous occurrence of major floods in the main stream of Nen River and the main stream of the Songhua River, peak flows at Harbin station were still relatively large.

To improve the flood control capacity of Harbin, several water conservation projects are planned for successive construction, including the Nierji Reservoir, Hadashan Reservoir, and reservoirs at the Gugu River, Liujiatun, Kuruqi, and Dalai. However, studies have shown that these reservoirs have minimal impact in reducing flood peaks in Harbin (Chen 2004).

In addition to the Baishan Reservoir and Fengman Reservoir, flood retention basins are an effective strategy in the event of a major flood. In the flood control system of Harbin, there are currently two major flood retention basins, namely the Pangtoupao retention basin and the Yueliangpao retention basin. When the Nen River and Songhua River were flooded in 1998, there were two levee breaches, with a total width of 538 m, at Pangtoupao. The peak discharge at the levee breach was 5040 m^3/s, with an average daily maximum flow of 4710 m^3/s. The total overflow of the breach was 6.435×10^9 m^3, and the volume of retained water was 3.53×10^9 m^3, submerging a total area of 1160 km^2, with an average depth of 3.05 m. The two retention basins fulfilled their role and reduced the flood peak flow below Dalai, greatly alleviating the flood control challenge in Harbin.

The two major flood retention basins, together with Baishan, Fengman, and Nierji reservoirs, jointly undertake the task of preventing massive floods that occur once every 100–200 years in Harbin, (Hong et al. 2013; Zhang 2016). However, due to the large number of residents, farmlands, oilfields, railways, and oil pipelines in the area, losses arising from flood retention are huge, and the retention basins cannot be used easily.

5 Analysis of Rainfall Characteristics in Recent Years

The above analysis of the effect of major flood control structures under different historical typical conditions demonstrate that the current flood control challenge remains significant for Harbin. In addition, as global climate changes, extreme precipitation events have occurred frequently in recent years, attracting extensive attention from scholars (Jiang et al. 2007; Yang et al. 2008; Zhai et al. 2007). In order to further understand the potential risks and clarify the needs for future flood control, this paper used rainfall data of Nen River and Songhua River upstream of Harbin, and at the Harbin station, to analyze the regional changes of rainfall characteristics in recent years, and discusses the possibility of future occurrence of extreme rainfall in Harbin.

Rainfall intensity at the Qiqihar station and the Tailai station in the upper Nen River Basin, and between the Qianguo station and Huadian station in the Second Songhua River Basin in the flood season, increased significantly after 2010 at all four sites (Fig. 2, 3) Rain intensity measured after 2010 increased from 7.67 mm/h, 6.89 mm/h, 7.38 mm/h, and 9.61 mm/h, to 8.42 mm/h, 7.9 mm/h, 7.88 mm/h, and 10.12 mm/h, corresponding to an increase of 9.77, 14.66, 6.78, and 5.30%, respectively, at the Qiqihar station, Tailai station, Qianguo station, and the Huadian station, during the flood season. The increase of rain intensity in the upstream poses a greater flood control challenge to Harbin. The number of heavy rainfall events in

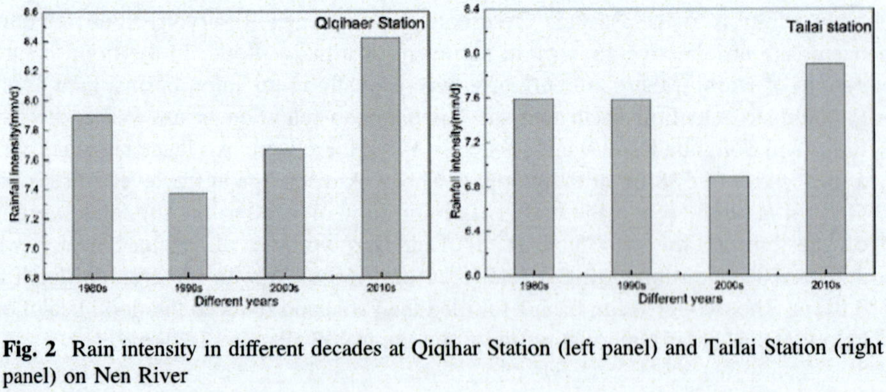

Fig. 2 Rain intensity in different decades at Qiqihar Station (left panel) and Tailai Station (right panel) on Nen River

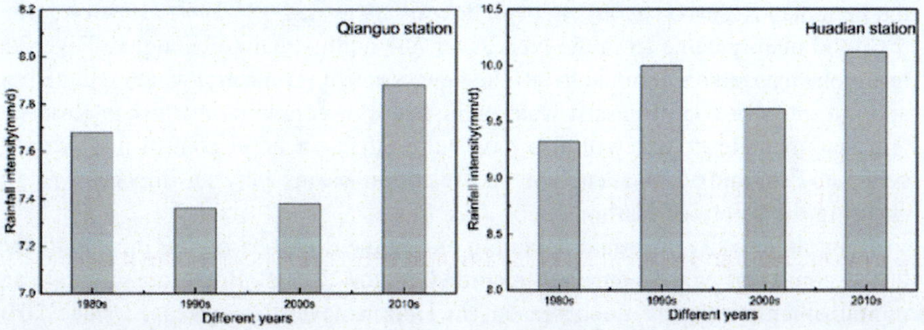

Fig. 3 Rain intensity in different decades at Qianguo Station (left panel) and Huadian Station (right panel) on Songhua River

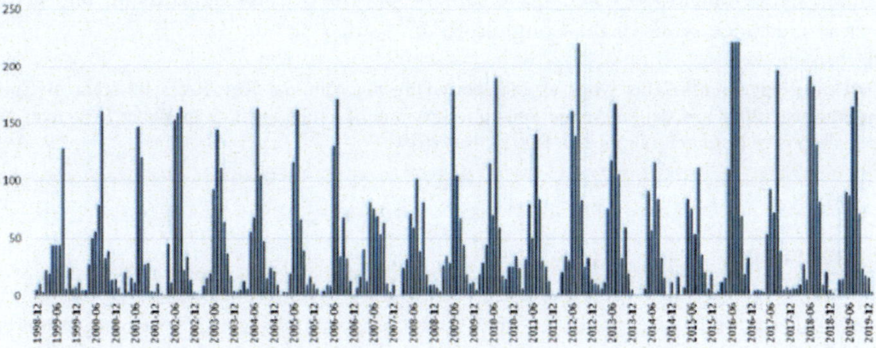

Fig. 4 Monthly cumulative rainfall in Harbin

Harbin during the flood season has not decreased in recent years, while rainfall intensity has increased (Fig. 4).

Thus, the flood control challenge in Harbin is increasing compared with earlier years. At the same time, because there are many river-related buildings and the flood retention sections are narrow in Harbin, the difficulty of flood control in Harbin is further increased.

6 Conclusions

This paper quantitatively analyzed the effects of the major flood control structures of Harbin under different typical historical flood conditions. The results demonstrated that the flood control effects of Baishan Reservoir and Fengman Reservoir varied according to the different types of floods. When the flood mainly came from the Second Songhua River Basin, the two reservoirs played an important role, with the Fengman Reservoir reducing peak flow by an average 33.19%. However, when the flood mainly came from the Nen River, the main stream of Songhua River, the reservoirs exerted a minimal effect. The new series of reservoirs, which will be built and put into use soon to assist water adjustment in conjunction Baishan Reservoir and Fengman Reservoir, will also have little effect on flood control in Harbin. In contrast, Pangtoupao and Yueliangpao retention basins play an important role in ensuring the safety of Harbin.

Rainfall intensity upstream of Harbin during the flood season has increased since 2010, compared with intensities recorded in the 2000s. Both total rainfall and rainfall intensity during flood seasons in Harbin have also increased since 2010. Therefore, there is a high possibility of extreme rainfall events in the future, and the aforementioned flood control structures alone cannot ensure the safety of Harbin. Consequently, the following actions are required to further ensure the flood control safety in Harbin: improve the joint use of flood control structures, urban embankment dams, and flood discharge channels; optimize water adjustment; and accelerate clearance work in the Songhua River.

Acknowledgements This work was supported by the National Key R&D Program of China (Grant No. 2018YFC0407305) and Scientific Program of China institute of Water Resources and Hydropower Research (Grant No. SE0145B792017).

References

HU Baoyuan, PAN Feng. Discussion on countermeasures for harmonious development of urban flood control and urban construction in Harbin. Water Conservancy Science and Technology and Economy, 2010, 16(6): 656–657.

WANG Zhixing, ZHANG Xiaoping, WANG Tianwei. Analysis and Countermeasures against Problems about Urban Flood Control for Main Urban Area in Harbin City. Heilongjiang Science and Technology of Water Conservancy, 2016, 44(4):106–109.

WU Dongping. Comprehensive evaluation of the construction of flood control system in Harbin. Harbin Engineering University, 2004.

MA Zhen. Basic Condition of Flood Control and Countermeasure and Measure for Precaution against Natural Calamities and Reduction of Natural Calamities in Harbin City. Urban Roads, Bridges, and Flood Control, 2005(3):5–8.

ZHAO Shuqing. Study on flood control scheme in Harbin City. Water Resources & Hydropower of Northeast China, 2006, 24(6):56–58.

MENG Xiangjuan, ZANG Jialun. Analysis of the problems and overall goals of the flood control projects in Harbin. Water Conservancy Science and Technology and Economy, 2011, 17(5): 75–75.

ZHANG Guifang. The status and role of the Pangtoupao flood detention basin in the flood control system of Harbin. Heilongjiang Water Conservancy Science and Technology, 2016, 44(7): 177–178.

CHEN Shi. Analysis of flood control effect and dam break effect of upstream reservoirs in Harbin. Water Conservancy Science and Technology and Economy, 2004, 10(4):230–230.

HAN Youbang. Suggestions on flood control in Harbin. Heilongjiang Water Resources, 2007(1): 42–42.

YUAN Guochen, JIN Mingqin. Reflections on flood control in Harbin in 1991. Northeast Water Resources and Hydropower, 1993(5):10–12.

JIANG Zhihong, DING Yuguo, CHEN Weilin. Projection of Precipitation Extremes for the 21st Century over China. Advances in Climate Change Research, 2007, 03(04).

YANG Jinhu, JIANG Zhihong, WANG Pengxiang, et al. Temporal and Spatial Characteristic of Extreme Precipitation Event in China. Climatic and Environmental Research, 2008(01):77–85.

ZHAI Panmao, WANG Cuicui, LI Wei. A Review on Study of Change in Precipitation Extremes. Advances in Climate Change Research, 2007, 003(003):144–148.

Evaluation Report on the Status of Harbin Urban Flood Control System. Heilongjiang Water Conservancy and Hydropower Survey and Design Institute, 2019.

HONG Wenbin, ZANG Yongshun, JIANG Pan, GUAN Chuantao. Analysis on the operation method of Pangtoupao and Yueliangpao flood detention basins. Northeast Water Resources and Hydropower, 2013(4):52–53.

ZHANG Guifang. The status and role of Pangtoupao flood detention basin in the flood control system of Harbin. Heilongjiang Water Resources Science and Technology, 2016, 44(7): 177–178.

Global Bibliometric Analysis of Research Activities on Plant Defense Against Abiotic Stresses from the Web of Science (2005–2017)

Jian Zheng, Zhengjiang Feng, Chuanyuan Zhu, Xingyun Qi, Yan Wang, and Jian Wang

Abstract Agriculture is fundamental for food production. The aim of modern agriculture is to improve or maintain food quality and production without exerting any negative influence on the environment. However, crops grow in a complex environment, and they suffer adverse effects from external factors. Non-biological factors (e.g., drought, low temperature, high temperature, heavy metals) can cause abiotic stress in plants, resulting in physiological injuries in plants and eventually leading to reduced plant productivity. To combat abiotic stress, plants have developed complex defense mechanisms in their growing stages. Therefore, understanding these defense mechanisms can help prevent crop yield losses. In this study, a bibliometric analysis of abiotic stress was conducted based on the Web of Science database. Different aspects of publications, including publication type, research field, journal type, country and its institutions, frequencies of keyword occurrence, and plant species in abiotic stress research, were analyzed. Results showed that abiotic stress research is flourishing worldwide, and the number of publications is increasing yearly. The journals selected by authors to publish their outputs are all related to plants, with the Journal of Experimental Botany being the most influential. From a global perspective, countries are in close cooperation on abiotic stress research. According to the results of keyword analysis, rice, wheat, tobacco, maize, and tomato are the most studied plant species. And drought, high salt, active oxygen, and heavy metals are found to induce abiotic stress more than

J. Zheng (✉) · Z. Feng · C. Zhu · X. Qi · Y. Wang
College of Energy and Power Engineering, Lanzhou University of Technology, 730050 Lan Zhou, P. R. China
e-mail: zhj16822@126.com

J. Zheng · Z. Feng · C. Zhu · X. Qi
Key Laboratory of Complementary Energy System of Biomass and Solar Energy, 730050 Lan Zhou, Gansu Province, P. R. China

J. Zheng · Z. Feng · C. Zhu · X. Qi
Western Energy and Environment Research Center, Lanzhou University of Technology, 730050 Lan Zhou, P. R. China

J. Wang
YangLing RuiWo Wr&HP Planning and Design CO. LTD., Yangling, P. R. China

R. G. Gutierrez (ed.), *Wastewater Technologies and Environmental Treatment*, Springer Proceedings in Earth and Environmental Sciences,
https://doi.org/10.1007/978-3-030-61989-3_4

other factors. In conclusion, sustained efforts have been made by the scientific community in the field of abiotic stress with the purpose of understanding, regulating, and controlling the negative effects of abiotic stress on plants.

Keywords Abiotic stress · Agriculture · Bibliometric analysis · Plant species · Publication

1 Introduction

Agricultural production is facing the enormous challenge of providing enough food for the continuously growing population worldwide with limited natural resources. To meet this challenge, the support of agriculture and its related sciences will be needed, and theories and techniques applied in agricultural systems need to be understood. The production of high-quality crops under various constraints without any negative effects on the environment is the major aim of modern agriculture (Gázquez 2016) and is important in sustainable agriculture. Constraints that come from adverse environmental conditions, such as drought, temperature, salt, and heavy metals, are called abiotic stresses. They break the normal physiological balance of plants in their natural growth process (Jaleel 2009) and are the main factors that limit agricultural productivity. Plants can tolerate a variety of abiotic stresses through a series of physiological and biochemical reactions. On the one hand, plants have evolved complex defense networks, also known as innate immune systems, in response to changing environmental conditions (Nejat 2017). On the other hand, drought, salt, and cold stresses can induce genes to react to stress at the transcription level, and these genes play great roles in stress response and tolerance (Gimenez 2018). Therefore, understanding the defense mechanism of plants and taking advantage of it can allow us to develop strategies to decrease crop yield reduction due to abiotic stress. Research efforts are focused on developing measures to make crops more resilient to mitigate the effects of abiotic stress on crop production (Balestrini et al. 2018), and numerous works have been published with the aim to understand the plant defense response to abiotic stress (Campos 2018; Gázquez 2016; Haines 2003) and improve current techniques in crop cultivations.

To understand the defense mechanism and adaptation of plants in response to adverse environmental conditions, numerous studies have been conducted since 2005 (Karkanis 2019; El-Shikha 2008). However, the current state of research; the contributions in the knowledge field of different countries; and the research institutions, researchers, and their major focuses are still not known. Fortunately, scientometric, informetric, and bibliometric sciences can be used to evaluate the scientific trends in some research fields (Gimenez 2018; Cañas-Guerrero 2013). The modes and trends of scientific publications can be quantitatively summarized with mathematical and statistical methods. Moreover, the citation, content, and academic achievements of institutions and countries can be evaluated to determine the development model and time evolution process of research in this field (Rojas-Sola 2015). This analysis was considered as a socioeconomic indicator of countries and research institutions because scientific production is closely related to financial investment (Salmerón-Manzano 2017). Therefore, the present study conducted a bibliometric analysis on global publications in the research field of abiotic stress in plants based on the Web of Science database. A total of 16,094 research papers related to abiotic stress published in 2005–2017 were analyzed to reveal the spatial distribution of countries, institutions, source journals, and time evolutions of scientific production. The analysis can provide a reference for further studies on abiotic stress in plants.

2 Materials and Methods

This research was performed using bibliometric analysis. The Web of Science database (Indices: Sci-Expanded, Cpci-S, Ccr-Expanded, Ic.) was used to extract scientific publications with the aim of compiling academic articles, including specific topic "abiotic* stress*" in the title, abstract, and/or keywords. Literature was restricted to the period between 1986 and 2017, and the research was conducted in March 2018. Truncation symbols ("*") in search topics were used to collect all documents related to abiotic stress. A total of 16,094 document index information closely related to the research topic were identified and studied. The technical publications obtained from the search were assessed and classified according to the following aspects: number of scientific publications for each year, type of document, journals, subject classification, country, and research institution.

The frequency of keywords in scientific publications has been commonly used to determine research trends in a specific field. In the present study, the keyword text file about abiotic stress is formatted using the Vos viewer software, and WordArt was used to define the display size of each word and draw the word cloud according to keyword frequency. The relative differences and relevance among the frequencies of keywords were analyzed using word cloud.

The academic, data, and application values and the influence of scientific literature in the scientific community were evaluated by the number of cited times. Therefore, the number of citations was used as an analytical tool, and the impacts of the relevant documents in the field of abiotic stress was analyzed using the h-index. The h-index is defined as h of a researcher's total articles has at least been cited h times each (Hirsch 2014). It was used to study the scientific contributions of a journal, country, and institution (Bornmann 2011). Important research results always have a high h-index, and evaluation using h-index has been successfully applied in previous bibliometric studies (Li 2018; Montoya 2014).

3 Results and Discussion

3.1 *Development of Scientific Output and Distribution in Subject Categories*

Within the retrieval time from 2005 to 2017, a total of 16,094 scientific publications were found in the "abiotic stress" research field. As shown in Fig. 1, the number of publications has grown significantly in recent years. Before 2007, the articles on abiotic stress were below 100, and this number has prominently increased to 609 in 2008, which is approximately 10 times that in 2007. The annual publications on abiotic stress research have grown continuously, and the number of articles is 1,146 in 2011 and has increased to 2,313 in 2015. Therefore, the number of publications increased in three periods: 2008, 2011, and 2015. The development illustrates that the scientific community has a strong interest in abiotic stress. Thus, a prediction can be made that more research on abiotic stress will be conducted in the coming years.

The development trend illustrates that the scientific community has paid close attention to abiotic stress research in recent years. A constant annual increase of published articles on abiotic stress since 2008 can be observed, and the year 2008 can be considered as a time demarcation point for research activity in this field. A study on the subject area distribution of publications was conducted to understand this unexpected rise. Results showed that "Plant Science" is the subject area with the highest number of publications on abiotic stress with 1,395 publications related to this field by 2017. The second area is "Biochemistry Molecular Biology," followed by "Agriculture" with 350 publications, as shown in Fig. 2. This outcome may have resulted from the close attention of the scientific community to new techniques in agricultural production, which allows for considering the negative influences of environmental conditions where plants grow, with the aim of improving irrigation and cultivation management.

Fig. 1 Trends in publications in the field of abiotic stress research in 2005–2017

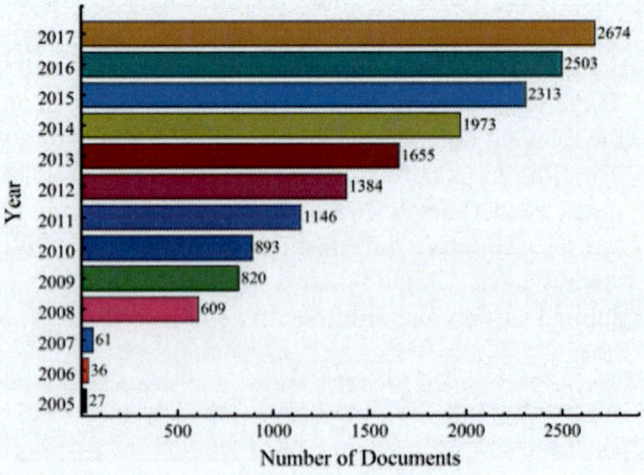

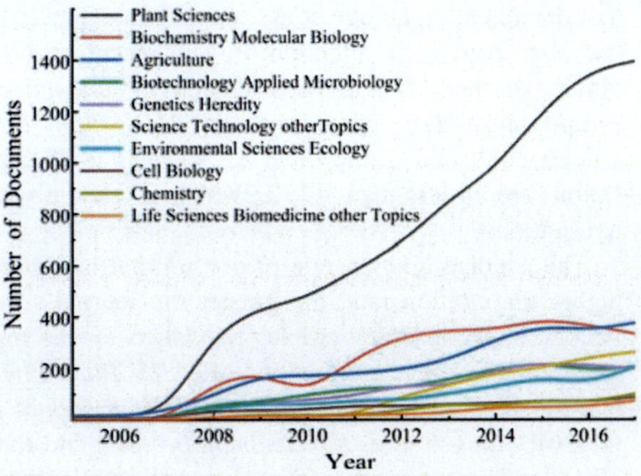

Fig. 2 Time evolution of the distribution of publications on abiotic stress research by subject area

3.2 *Types and Distributions of Output in Publications*

The documents on abiotic stress retrieved from the Web of Science database are classified into 17 document types, and Table 1 lists the first five types of the publications. The "Articles" type accounts for 83.3% of the total publications with 13,411 records and is the most common document type in abiotic stress research. The second place is "Reviews," accounting for 11.7% with 1,780 records, followed by "Proceedings Paper" with 591 reports, accounting for 3.7%. "Meeting Abstract" has 106 documents, accounting for 0.7%. These four types of documents account for 98.8% of the total publications from abiotic stress research. "Book Chapter," "Editorial Material," "Letter," and "Book Review" added up to 1.2%. The results indicated that articles are the first choice of most authors in publishing their scientific outputs.

Taking the sources into consideration, the results showed that more than 1,000 journals have published research documents on abiotic stress according to the classification of journals in Web of Science. A total of 137 journals were observed with more than 20 documents related to abiotic stress, and the 20 most productive journals are listed in Table 2. Although the 20 journals account for only 2% of the total number of journals, nearly 40% of the research documents related to abiotic stress were published by them. *Frontiers in Plant Science* leads the list with 976 documents and a percentage of 6.06% of all publications on abiotic stress, and its

Table 1 Distributions of research on abiotic stress

Document Types	Number	% of 16,094
Article	13,411	83.3
Review	1,780	11.1
Proceedings paper	591	3.7
Meeting abstract	106	0.7
Others	206	1.2

first document on abiotic stress was published in 2011, demonstrating that this field had been paid great attention only in recent years. The second is the multidisciplinary journal *PLoS ONE*, followed by *Journal of Experimental Botany*, *Plant Physiology and Biochemistry*, and *BMC Plant Biology*. The largest number of articles published in the other 15 journals is 281, and their proportions of publications are all less than 1.75%, which is relatively small compared with the above five journals.

The average citation rate of one journal is a direct indicator of its impacts: the higher the citation rate, the greater the journal's impact in this field (Liu 2011). Abiotic stress research articles published on the top 20 most productive journals received an average citation rate of 22.75. *Plant Physiology* has the highest citation rate (49.17), followed by *Journal of Experimental Botany* (43.69); *Plant Journal* (43.26); *Plant, Cell & Environment* (40.81); and *Plant Molecular Biology* (31.23). The citation rates in other journals are all smaller than 27. Interestingly, *Frontiers in Plant Science* has the largest number of publications on abiotic stress research, but its citation rate is only 9.71, ranking in the last three journals. This finding may be because its first published document on abiotic stress was in 2011, a relatively late publishing in this field (Fig. 1).

The h-index is also considered as an important indicator in analyzing a journal's influence. From this point of view, the most influential journals were analyzed, and results are shown in Table 2. *Journal of Experimental Botany; Plant Physiology; Plant Journal; Plant, Cell & Environment;* and *BMC Plant Biology* cover the top five out of the 20 influential journals, and their h-indices are 77, 71, 51, 49, and 49, respectively.

In this section, the top 20 source journals with relatively high influence and publishing capability in the field of abiotic stress research were identified, allowing us to capture valuable journal objects from literature, which is convenient for further detailed analysis. Combining the total number of publications, number of times cited, and h-index, the most valuable journal related to abiotic stress research was the *Journal of Experimental Botany* as it was ranked in the top five of all indicators out of the 20 source journals.

3.3 Publication Distribution by Countries and Institutions

Publication Distribution by Countries

According to the 2017 statistics, abiotic stress has been researched by 145 countries, and these countries are distributed in almost every continent, proving that worldwide interest has been aroused on the research of abiotic stress in plants (Fig. 3). China (4,972 publications, 30.9%), the USA (2,430 publications, 15.1%), India (1,764 publications, 11.0%), South Korea (910 publications, 5.7%), Germany (908 publications, 5.6%), Spain (829 publications, 5.2%), Japan (690 publications, 4.3%), France (679 publications, 4.2%), and Italy (650 publications, 4.0%) have published more than 86% of the total number of documents about abiotic stress in

Table 2 Distribution of publications by source

Rank	Journal	TP[a]	% of 16,094	ATC[b]	ALCR[c]	H-index	Country
1	Frontiers in plant science	976	6.1	9.71	8.12	48	Switzerland
2	PLoS one	684	4.3	14.74	5.49	47	USA
3	Journal of experimental botany	403	2.5	43.69	5.33	77	UK
4	Plant physiology and biochemistry	374	2.3	21.50	4.48	41	France
5	BMC plant biology	319	2.0	23.44	5.16	49	England
6	Plant physiology	281	1.7	49.17	4.54	71	USA
7	Plant cell reports	273	1.7	18.18	6.07	37	Germany
8	Journal of plant physiology	272	1.7	16.99	4.05	40	Germany
9	Acta physiologiae plantarum	264	1.6	7.46	5.09	24	Germany
10	Plant science	256	1.6	18.21	5.39	38	Ireland
11	Scientific reports	249	1.5	5.30	6.45	22	England
12	BMC genomics	247	1.5	21.06	5.20	38	UK
13	Plant molecular biology	224	1.4	31.23	5.22	45	Netherlands
14	Planta	218	1.4	26.41	4.51	41	Germany
15	International journal of molecular sciences	187	1.2	17.99	7.66	32	USA
16	Plant, cell and environment	182	1.1	40.81	5.79	49	England
17	Plant molecular biology reporter	181	1.1	9.66	5.52	23	USA
18	Environmental and experimental botany	173	1.1	20.61	5.55	35	England
19	Plant journal	173	1.1	43.26	4.06	51	England
20	Molecular biology reports	166	1.0	15.50	3.83	27	Netherlands

[a]TP: Total number of publications; TC: Total cited score; LCR: Local cited references
[b]ATC = TC/TP
[c]ALCR = LCR/TP

the period analyzed, and they belong to the top 10 countries based on the number of publications (Fig. 3). Five of the top 10 countries are located in Asia, and none belong to Africa, which is the region with the smallest number of publications on abiotic stress and wherein most countries have a publication number of less than 30.

Figure 4 shows the full cooperation network among 31 countries in abiotic stress research. The link width among countries presents the amount and intensity of their international collaboration. The thicker the line between two countries, the stronger the mutual collaboration between them. For example, the countries with the closest cooperation with the USA in abiotic stress research are China, South Korea, Spain,

Fig. 3 Worldwide research on abiotic stresses in 2005–2017

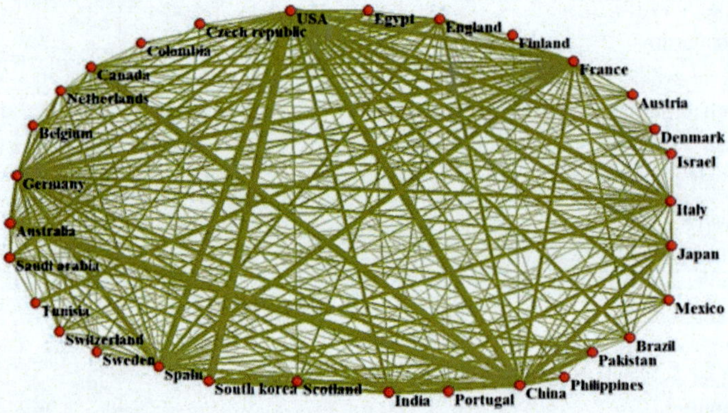

Fig. 4 Full cooperation network between different countries in the field of abiotic stress

and France according to the thickness of the line in the network. China, the USA, France, Germany, and Spain are the countries that have the closest cooperation with other countries. In addition, a relatively close collaboration can be found among other countries, which is further proof that abiotic stress research has attracted worldwide attention and indicates mutual cooperation aimed at jointly addressing the impact of abiotic stress on plant growth and production and people's life in a global scale. The results also showed that abiotic stress research in each country is focused on plant species (Fig. 5). The most studied plant is *Arabidopsis*, which has a short life cycle, small genome, and accessible transformation methods and is considered as model species in abiotic stress research activities. China published approximately 3,500 documents related to abiotic stress in *Arabidopsis*, which is by far more than other species used in this field. Rice is the second most widely studied plant species in China, India, the USA, Korea, Japan, France, and Germany. The reasons may be that rice is one of the main food, and rice-eating population accounts for approximately 46% in the world (Roberts 2009), especially in China,

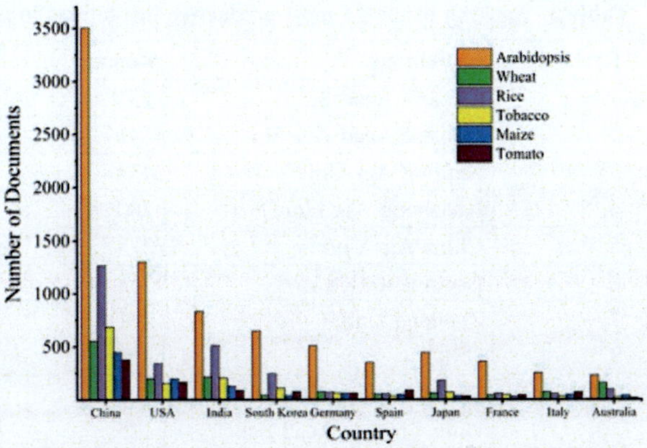

Fig. 5 Distribution by country of abiotic stress reports categorized by plant species

India, South Korea, and Japan where it is a staple food. Meanwhile, wheat is the second most widely studied plant species in Australia, which may be because it is a main export crop. Tobacco is another plant species used in research related to abiotic stress, whereas tomato is the least studied species except in Spain and Italy. Interestingly, China headed the top 10 list in terms of the number of publications on abiotic stress, and tomato is the least studied species in China.

Publication Distribution by Institutions
Research institutions constitute the research strength of a country. On the basis of the results mentioned above, the most productive institutions with many publications related to abiotic stress are from China and the USA. Table 3 presents the 14 most productive institutions that have more than 150 publications on abiotic stress during the analyzed period. Nine of these research institutions are from China, three from the USA, and the other two from Spain and France.

3.4 Languages Used in the Abiotic Stress Research Publications

Thirteen different languages have been used to publish studies in the field of abiotic stress research. As expected, English is the most widely used because it is an international language commonly used in the scientific community. A total of 15,972 documents out of 16,094 or 99.242% of publications are written in English. The second used language is Portuguese (41 reports, 0.225%), followed by Spanish (17 reports, 0.106%). Among them, more than 500 research papers have been published in Brazil, and more than 200 research papers have been published in Portugal (Fig. 4). Spain ranks seventh among the top 10 countries in the number of publications, which is the reason why Spanish is used more in abiotic stress research field. Compared with Spain, China is the country with the largest number of publications, but the use of Chinese language is limited in all publications

Table 3 Ranking of the 14 most productive institutions in abiotic research

Rank	Institutions	Reports	% of 16,094	Country
1	Chinese Acad Sci	751	4.7	China
2	Chinese Acad Agr Sci[a]	441	2.7	China
3	Nanjing Agr Univ[a]	434	2.7	China
4	Huazhong Agr Univ[a]	280	1.7	China
5	China Agr Univ[a]	273	1.7	China
6	Northwest a f Univ[a]	258	1.6	China
7	Zhejiang Univ	251	1.6	China
8	CSIC	217	1.3	Spanish
9	Shandong Agr Univ[a]	203	1.3	China
10	Univ Chinese Acad Sci	175	1.1	China
11	Inra[a]	165	1	France
12	Univ Calif Davis[a]	159	1	USA
13	Usda Ars[a]	159	1	USA
14	Ars[a]	154	1	USA

[a]Research Institutions focused on Agronomy

Table 4 Languages of scientific output in the abiotic stress research field

Rank	Language	Reports	% of 16,094
1	English	15,972	99.242
2	Portuguese	41	0.255
3	Spanish	17	0.106
4	Polish	16	0.099
5	Czech	12	0.075
6	French	11	0.068
7	Chinese	8	0.050
8	Korean	5	0.031
9	German	4	0.025
10	Turkish	4	0.025
11	Russian	2	0.012
12	Croatian	1	0.006
13	Slovak	1	0.006

(Table 4), which may be because most journals included in Web of Science need to be written in English. As shown in Table 2, the 20 journal editorial departments with the largest number of publications are all located in Europe and the Americas; none of them are located in Asia.

Analysis of Keywords

The topic of a research manuscript can be determined by keywords. Thus, if all keywords are summarized, analyzed, and evaluated, the research trend in a specific field can be obtained (Gimenez et al. 2017). In this paper, a total of 42,530 keywords used in publications related to abiotic stress were collected, and the words with an occurrence of greater than 50 times were selected for present analysis. After screening, 516 keywords, which account for nearly 1% of all the keywords, were obtained. Their total occurrence frequency is 103,192, accounting for 48% of the total frequency of all keywords (214,816). This finding is proof that the 516 keywords selected are the core words in abiotic stress research. For further study, we combined the words such as “stress” and “stresses” and “crop” and “plant” and removed the word “abiotic stress” as it was used to retrieve the database. Finally, a total of 282 keywords were obtained, and a word cloud was made using these keywords (Fig. 6).

As seen in Fig. 6, the first type of keywords that appear more frequently in the field of abiotic stress is related to plant species, including *Arabidopsis*, rice, wheat, tobacco, maize, and tomato. In fact, 18 other plants, such as soybean and *Medicago truncatula*, with lower frequency of occurrence are also found in the keyword analysis. Most of these plants are important crops in agricultural production.

The second most frequent keyword type is phytohormones. “Abscisic acid” (ABA) occurred more than other phytohormones in abiotic stress research. ABA is a very important plant hormone involved in many plant stress behaviors caused by abiotic stress and controls the transcription of plant tolerance genes (Cramer 2011). “Proline,” which is a controversial plant hormone in plant tolerance, is the second type of phytohormone. Some scholars believe that its accumulation in plants is a response to osmotic stress, whereas others consider that it is a kind of permeate and reactive oxygen species (ROS) scavenger, which can be used to prevent damage caused by high salt and drought stresses (Natarajan 2012; Szabados 2010). It appeared as a keyword many times in the bibliometric analysis of abiotic stress, indicating that it is one of the most critical plant hormones in the field of abiotic

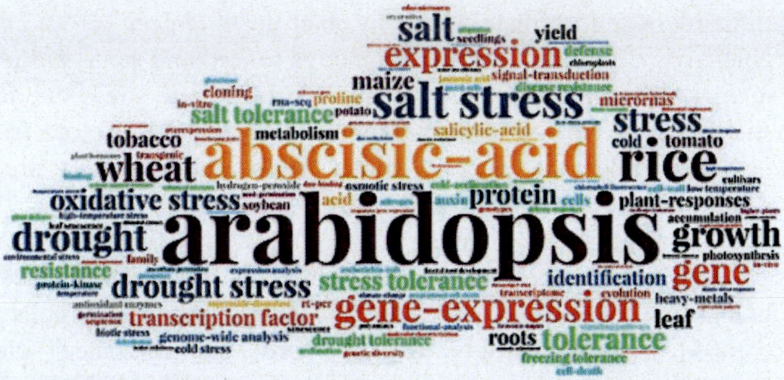

Fig. 6 Word cloud based on the main keywords related to abiotic stress worldwide research

stress. The third plant phytohormone is "salicylic acid," which is an important regulatory signal for plant response to abiotic stresses, such as drought, cold, and high temperatures (Waterman 2019). "Hydrogen peroxide," "nitric oxide" (NO), and "antioxidants" have appeared repeatedly in the keyword analysis of abiotic stress. Previous studies have shown that abiotic and biotic stresses can induce plants to synthesize hydrogen peroxide and NO, which can mediate the transcription of plant specific genes and affect the stomata closure of plant (She 2004). NO and ROS have been shown to synergistically increase the death of host cells in soybean suspension cells (Delledonne 1998), and antioxidants are important chemicals in plants for dealing with oxidative stress.

The third most frequent keywords, which include "gene-expression," "expression," "gene," "signal-transduction," "microRNAs," and "cloning," are related to plant genetics. Plant growth and development result from the comprehensive expression of genes in time and space through various metabolic and physiological processes. Stress tolerance is the level of adversity factor stresses that plants can bear. Improving crop tolerance can be regarded as a fundamental measure to reduce the adverse effects of environmental factors on growth and development. Therefore, the development and isolation of potentially valuable genes and intense research on its expression mechanism are the keys to improve crop varieties and tolerance (Van Oosten 2014). With the development of biotechnology, full genome sequencing of many plants allowed the cloning of genes, laying the foundation for studying the mechanism of plant stress resistance at the molecular level. At the same time, this field also provides a new way to take genetic research as an entry point to study plant growth and development regulation, stress and defense response, and plant cell apoptosis. In this study, gene-related keywords have become the main frequently used keywords in abiotic stress research analysis, revealing the wide application of gene and biotechnology in this field from another aspect.

The fourth most frequent keywords are related to stress and plant tolerance. Because improving crop tolerance is the aim of abiotic stress studies, the keywords involved in stress and crop tolerance were analyzed together in this study. "Drought stress" and "salt stress" are the first types of stress highlighted in Fig. 6. Drought stress can lead to a decrease in plant water content and leaf water potential and leaf stomata closure. Severe water stress may cause photosynthesis to cease and metabolic disorders and ultimately lead to plant death (Jaleel 2009). Drought is a global problem as one-third of the Earth belongs to arid and semi-arid regions, and drought stress has become the most severe type of abiotic stress due to its widespread influence. Moreover, "drought tolerance" has appeared more frequently in keywords of plant tolerance. Second, under the condition of salt stress, a large amount of salt ions enters the plant and accumulates, which will break the original balance of free radicals in the plant and change the composition, permeability, and transport capacity of plant cell membranes, damaging the function and structure of cell membranes and seriously affecting the normal function of plant cells and tissues (Netondo 2004). Correspondingly, the keyword "salt tolerance" can also be observed in the word cloud. The second type of stress highlighted is "oxidative stress." ROS are produced as by-products of various metabolic pathways in different cellular organs, such as chloroplasts, mitochondrion, and peroxisomes (Navrot 2010; Gill 2010). Normally, active oxygen is removed by the plant's antioxidant mechanism; however, abiotic stresses, such as salinity, ultraviolet radiation, drought, heavy metals, and extreme temperatures, can lead to excessive production of ROS in plants. ROS is highly reactive and toxic; it can cause damage to proteins, lipids, carbohydrates, and DNA and ultimately results in oxidative stress (Gill 2010). Because oxidative stress can be caused by other abiotic stresses and is easily affected by many environmental factors, it attracted more attention compared with other stresses. Of note, no keyword specific to oxidative tolerance can be observed in the word cloud because increasing plant tolerance to other stresses is a means of solving plant oxidative stress. "Osmotic stress" is the third type of stress of great concern in abiotic stress research. It generally refers to the situation of water shortage that limits plant growth and development. Drought, excessive salinity, low temperature, and freeze can cause osmotic stress in plants. The other two types of stress that received much attention in the field of abiotic stress research are both caused by temperature: "high-temperature stress" and "cold stress." Temperature is an important factor that limits plant growth and spatial distribution. High temperature stress can induce significant changes in photosynthesis, dark respiration, membrane stability, and mitochondrial respiration, resulting in crop yield reduction (Moaed 2006). Cold stress includes chilling injury (temperature below 20 °C) and freezing injury (temperature below 0 °C). Low temperature can cause flower abortion, pollen and ovule infertility, and fertilization obstacles and affects seed filling, ultimately resulting in low seed setting rate and crop yield reduction (Thakur 2010). "Freezing tolerance" also appeared many times in keywords about plant tolerance. Moreover, another type of stress that cannot be ignored in the word cloud is "heavy metal stress." Heavy metals, such as Hg, Cd, Pb, Cr, Cu, and Zn, can be derived from water, soil, and even toxic gases. Although Cu and Zn are necessary metal elements for plant growth, excessive metal content has adverse effects on plant growth (Rai 2019). With the development of modern industry and use of chemical fertilizers, heavy metal pollution may increase if not managed properly, and excessive heavy metals may accumulate in plant organs and enter the food chain, causing great harm to humans and animals. The last keyword that appeared in the word cloud is "biotic stress," which is because various abiotic and biotic factors always act in combination in plants. As such, the effects of the two factors should be analyzed together in the perspective of plant growth.

At the same time, words that describe plant physiological activities, such as "growth," "yield," "leaf," "roots," "metabolism," "photosynthesis," and "chloroplasts," appear in the keyword analysis. Growth and yield are the direct results of various metabolism and physiological activities. Metabolism refers to various chemical processes involved in the survival of organisms. These chemical processes change when plants are subjected to abiotic stress, finally causing physiological damage to plant. Photosynthesis is an important physiological process for plants to accumulate dry matter. Leaves are places for photosynthesis and gas exchange in plants. Their degree of opening and conduction of stomata greatly affect photosynthesis. Many abiotic stresses can cause stomatal closure, negatively affecting plant photosynthesis and normal life activities. The roots are organs for plants to absorb water and nutrients, and drought stress, salt stress, and heavy metal pollution affect plants through the roots. Thus, the influence of abiotic stress on root growth also affects the growth and yield of crops through various metabolic and physiological activities of plants. Yield is the main goal pursued by people in agricultural production, and plants suffer various stresses throughout their life cycle, so studying abiotic stress is beneficial to improve the ability to grow well in adversity.

4 Conclusions

Agricultural products, such as rice, wheat, maize, vegetables, and fruits, are essential food for people on a global scale. Plants provide the material base for human existence. However, they grow in the natural environment, and they are inevitably exposed to some adverse environmental conditions and suffer abiotic stress from drought, high temperature, cold, salt, and heavy metals, causing serious agricultural and economic losses. Therefore, abiotic stress on plants has attracted extensive attention and has been studied worldwide. In the present study, researches on abiotic stress were analyzed from the view of spatial distribution of countries, institutions, research documents, source journals, and time evolution of publication number. Moreover, research trends on the abiotic stress were evaluated and summarized by keyword analysis using bibliometric methods. The results showed that abiotic stress research is flourishing worldwide as 145 countries have expressed keen interest on abiotic stress studies, and scientific research centers of these countries communicate and cooperate with each other. From the time evolution of abiotic stress research, the scientific community paid increasing attention to this field, and the number of publications increased every year. For the research concerns, stress response and metabolism changes of plants when they suffer abiotic factors, especially the effects of abiotic stress on plant growth and yield, and genetic characteristics are mostly studied. *Arabidopsis* is commonly used as a model plant, and crops such as rice, wheat, maize, tobacco, and tomato are also deeply studied. Through keyword analysis, drought, high salt, active oxygen, and heavy metals are found to induce abiotic stress more than other factors. In conclusion, the results indicated that the scientific community has made great efforts in abiotic stress research with the aim of understanding, regulating, and controlling the adverse effects of abiotic stress to plants. These results can help us understand the development trend of abiotic stress research.

Acknowledgements The authors would like to acknowledge the financial support given by the National Natural Science of China (51969012) and Gansu Provincial Higher Education Science and Technology Achievements Transformation Project (2018D-04), Gansu Provincial Natural Science Foundation (18JR3RA154), Yangling Demonstration Zone Collaborative Innovation Major Project (2018CXY-14), Red Willlow First-class Discipline Project of LUT (0807J1) and Web of Science database for providing data.

References

Bornmann, L.: A multilevel meta-analysis of studies reporting correlations between the index and 37 different index variants. Journal of Informetrics 5(3), 346–359 (2011).

Balestrini, R.: Improvement of plant performance under water deficit with the employment of biological and chemical priming agents. Journal of Agricultural Science 156, 680–688 (2018).

Cañas-Guerrero, I.: Bibliometric analysis of research activity in the "Agronomy" category from the Web of Science, 1997–2011. European Journal of Agronomy 50, 19–28 (2013).

Cramer, G.R.: Effects of abiotic stress on plants: a systems biology perspective. *BMC Plant Biol.* 11, 163 (2011).

Campos, F.P.: Productivity and physiological response of lactating Holstein cows to sugarcane supplementation and evaporative cooling under heat-stress. Journal of Agricultural Science 156, 900–913 (2018).

Delledonne, M.: Nitric oxide functions as a signal in plant disease resistance. Nature 394, 585–588 (1998).

El-Shikha, D. M.: Remote sensing of cotton nitrogen status using the canopy chlorophyll content index (ccci). Transactions of the Asabe 51(1), 73–82 (2008).

Gázquez, J.A.: Intelligent low cost telecontrol system for agricultural vehicles in harmful environments. Journal of Cleaner Production 113(Feb.1), 204–215 (2016).

Gimenez, E.: DNA Damage Repair System in Plants: A Worldwide Research Update. *Genes*. 8 (11), 299–316 (2017).

Gimenez, E.: Worldwide Research on Plant Defense against Biotic Stresses as Improvement for Sustainable Agriculture. *Sustainability*. 10(2), 391 (2018).

Gill, S. S.: Reactive oxygen species and antioxidant machinery in abiotic stress tolerance in crop plants. Plant Physiology & Biochemistry 48(12), 909–930 (2010).

Haines, M.M.: Abnormalities in growth, development and physiological responses to biotic and abiotic stress in potato (Solanum tuberosum) transformed with Arabidopsis ETR1. Journal of agricultural science 141(3–4), 333–347 (2003).

Hirsch, J.E.: The meaning of the h-index. International Journal of Clinical and Health Psychology 14(2), 161–164 (2014).

Jaleel, C.A.: Drought stress in plants: a review on morphological characteristics and pigments composition. International Journal of Agriculture & Biology 11, 100–105 (2009).

Karkanis A., G.: Interference of weeds in vegetable crop cultivation, in the changing climate of Southern Europe with emphasis on drought and elevated temperatures: A review. The Journal of Agricultural Science 156(10), 1175–1185 (2019).

Li, X.P.: Consistent improvements in soil biochemical properties and crop yields by organic fertilization for above-ground (rapeseed) and below-ground (sweet potato) crops. Journal of Agricultural Science 156, 1186–1195 (2018).

Liu, X., L.: Global biodiversity research during 1900–2009: a bibliometric analysis. Biodiversity & Conservation. 20(4), 807–826 (2011).

Moaed, A.: Protective role of antioxidant enzymes under high temperature stress. Plant Science. 171(3), 382–388 (2006).

Montoya, F.G.: The research on energy in spain: A scientometric approach. Renewable & Sustainable Energy Reviews 29, 173–183 (2014).

Natarajan, S.K.: Proline dehydrogenase is essential for proline protection against hydrogen peroxide-induced cell death. Free Radical Biology & Medicine 53(5), 1181–1191 (2012).

Navrot, N.: Reactive oxygen species generation and antioxidant systems in plant mitochondria. Physiologia Plantarum 129(1), 185–195 (2010).

Nejat, N.: Plant Immune System: Crosstalk Between Responses to Biotic and Abiotic Stresses the Missing Link in Understanding Plant Defence. Current Issues in Molecular Biology 23, 1–15 (2017).

Netondo, G.W.: Sorghum and salinity: I. Response of growth, water relations, and ion accumulation to NaCl salinity. Crop Science 44(3), 797–805 (2004).

Rai, P.K.: Heavy metals in food crops: Health risks, fate, mechanisms, and management. Environment International 125, 365–385 (2019).

Roberts, M.J.: World Supply and Demand of Food Commodity Calories. *Am. J.Agr. Econ.* 91(5), 1235–1242 (2009).

Rojas-Sola, J.I.: Global Bibliometric Analysis of the 'Mining & Mineral Processing' Subject Category from the Web of Science (1997– 2012). Mineral Processing and Extractive Metallurgy Review 36(6), 349–369 (2015).

Salmerón-Manzano, E.: Worldwide scientific production indexed by Scopus on Labour Relations. Publications 5, 25 (2017).

She, X.P.: Role and Relationship of Nitric Oxide and Hydrogen Peroxide in Light/Dark-regulated Stomatal Movement in Vicia faba. Acta Botanica Sinica 46(11), 1292–1300 (2004).

Szabados, L.L.: Proline: a multifunctional amino acid. Trends in Plant Science 15(2), 89–97 (2010).

Thakur, P.: Cold stress effects on reproductive development in grain crops: An overview. Environmental & Experimental Botany 67(3), 429–443 (2010).

Van Oosten, M.J.: The Role of the Epigenome in Gene Expression Control and the Epimark Changes in Response to the Environment. Critical Reviews in Plant Sciences 33(1), 64–87 (2014).

Waterman, J. M.: Simulated Herbivory: The Key to Disentangling Plant Defence Responses. Trends in ecology & evolution 34, 447–458 (2019).

Inactivation of Bioaerosols in Dentist Clinic by Carbon Nanotube Discharge Plasma

Hsiao-Chien Huang, Ying-Fang Hsu, Shinhao Yang, Chi-Yu Chuang, and Wei-Ting Liu

Abstract This work aimed to apply carbon nanotube discharge plasma system to explore the feasibility of inactivating bacterial bioaerosols in dentist clinic environments. A dentist clinic operating environment in a medical center had been chosen as the testing indoor environment. The nanotube carbon plasma device with operating voltages of 6.0 kV was used for evaluating the bioaerosol inactivating ability of this plasma system. The results showed that bacterial bioaerosol concentration in the test environment is 815–1,518 CFU/m^3. When using the nanotube carbon discharge plasma system, the bacterial bioaerosol concentration would decrease 70% with 120 min system working. The overall space bioaerosol inactivating ability is about 70%. These results indicated that carbon nanotube discharge plasma system had the effective ability on airborne bacterial bioaerosols inactivating.

Keywords Carbon nanotube discharge plasma · Bacterial bioaerosols · Dentist clinic

1 Introduction

Recently there have been reports of airborne bacterial bioaerosol contamination distribution during dental practice (Timmerman et al. 2004; Walker and Marsh 2007; Rautemaa et al. 2006). It would increase the patient's risk of infection. Pervious study showed that airborne bacterial concentration would reach 4312 ± 324 CFU/m^3 during scaling treatment in the dentist clinic (Chuang et al. 2014). Therefore, the removal of airborne bacterial bioaerosols in the dentist clinic has become a very important issue.

H.-C. Huang · Y.-F. Hsu · S. Yang (✉) · W.-T. Liu
Environmental Sustainability Lab, CTBC Business School, Center for General Education, Tainan, Taiwan, R.O.C.
e-mail: shinhaoyang@ntu.edu.tw

C.-Y. Chuang
Department of Occupational Safety and Health, School of Safety and Health Sciences, Chang Jung Christian University, Tainan, Taiwan, R.O.C.

R. G. Gutierrez (ed.), *Wastewater Technologies and Environmental Treatment*, Springer Proceedings in Earth and Environmental Sciences,
https://doi.org/10.1007/978-3-030-61989-3_5

Recently, many indoor air controlling technologies had been adopted for inactivating bioaerosols. These cleaning equipments included electret filters (Yang et al. 2007; Yang and Lee 2005), ultraviolet germicidal irradiation (Lin and Li 2002), ozone (Li and Wang 2003), photocatalyst (Sunada et al. 1998; Lin and Li 2003; Yu et al. 2006), negative ion cleaner (Krueger and Reed 1976; Tanimura 1997), and membrane-less electrolyzed water (Chuang et al. 2012, 2013).

This work applied the carbon nanotube discharge plasma system for inactivating bacterial bioaerosols in dentist clinic operating environment. This carbon nanotube discharge plasma system is an novel nonthermal plasma for inactivating bioaerosols. Several researchers had demonstrated that the traditional nonthermal plasma could inactivate microorganisms effectively in liquid phase (Sato et al. 1996; Laroussi 2002, 2005; Locke et al. 2006; Gadri et al. 2000; Sladek and Stoffels 2005). Also, the traditional nonthermal plasma could inactivating indoor *E. coli* bioaerosols (Gallagher et al. 2007). Our pervious indicated carbon nanotube discharge plasma system could inactivate bioaerosols effectively (Yang et al. 2011; Huang et al. 2013). The microorganisms inactivation mechanisms of plasma technology are UVC and VUV radiation, diffusion of oxygen-containing free radicals or oxygen species, charged particles bombard cell wall, and local, periodic and short-term heating of microorganisms (Gallagher et al. 2007). Also, the advantage of the carbon nanotube is that it has excellent field electron emission ability. In the discharge plasma system, applying carbon nanotube as the discharge electrode could decrease the threshold operating voltage. According to these pervious studies, this novel discharge system had the lower staring voltage and ozone by-product. Thus, this work used carbon nanotube plasma system in indoor environments for bioaerosols inactivating. And the goal of this work is evaluating the inactivating ability of carbon nanotube discharge plasma system on bacterial bioaerosols in the dentist clinic.

2 Experimental Methods

2.1 Sampling Sites

In this study, a dentist clinic operating environment in a medical center had been chosen as the testing indoor environment. The size of testing environments is about $8 \times 10 \times 3.5\ m^3$.

2.2 Plasma System

This work used the carbon nanotube discharge plasma for inactivate bacterial bioaerosols. This plasma system has been sued in our pervious study (Yang et al. 2011).

The carbon nanotube electrodes were chosen as the discharge electrode in this plasma system. The configuration of the nanotube electrode is using the nanotube coated and adhered to the stainless steel electrodes with silver paste. The carbon nanotube selected in this work was multi-walled carbon nanotube (Muiti-Wall Carbon Nanotube, MWCNT) produced by the arc-discharge method (Arc-Discharge). The discharge voltage range of the power supply was 1.5–7.5 kV. In this work, the operating voltages of 6.0 kV was applied to evaluate the effectiveness of the plasma system's bioaerosol inactivating ability. The pumping flow rate was controlled by a fan. The operating condition of pumping flow rate is set to 0.78 Lpm.

2.3 *Bacterial Aerosol Sampling Method*

An Andersen single-stage impactor sampler (Andersen Samplers, Inc., Atlanta, GA) was applied for sampling the bacterial bioaerosol concentration. Tryptone soy agar (TSA) plates was used for bacterial culture. Sampling flow rate and time was setting on 28.3 LPM and 60 s. The incubation of TSA plates sample was conditioned at a temperature of 30 ± 1 °C for 48 ± 2 h. After incubation, manually count the colonies formed on the plate samples and convert them to the airborne bacterial concentration in CFU/m^3.

2.4 *Air Exchange Rate (ACH) Calculation*

The test process uses carbon dioxide standard gas to increase the carbon dioxide concentration in the environment to 2500–3500 ppm, and then a carbon dioxide detector (Q-trak, TSI Inc) was applied to measure the carbon dioxide concentration at different times. Also, ACH in the indoor environment could be estimated by using the first-order decay characteristics of CO_2.

2.5 *Decay Behavior of Bioaerosols*

This work used the decay coefficient to illustrate the behavior of bioaerosols being inactivated. This decay coefficient was calculating by using the following equations:

The decay curves of bioaerosol concentration were more difficult to compare than the

$$dC/dt = -kC, \quad (1)$$

$$C_t = C_0 \exp(-kt), k = k_a, \quad (2)$$

where C is the concentration of bacterial bioaerosol (CFU/m^3); C_0 and C_t are the initial concentration of the bioaerosol in the test and its concentration at time t, respectively (CFU/m^3); k is the bioaerosol decay coefficient (min^{-1}); k_a is the decay coefficients of bioaerosol associated with carbon nanotube discharge plasma system inactivating effects (min^{-1}); t is the testing time (min). In each test, the C_0, C_t and t would be sampled and recorded.

3 Results and Discussions

3.1 ACH of the Testing Room

This study used carbon dioxide standard gas to increase the carbon dioxide concentration in the testing environment to 2500–3500 ppm, and the carbon dioxide concentration at different times was measured to estimate the air exchange rate of the testing environment. Figure 1 shows the ventilation rate test of the dental clinic. During the test, four points in each area are selected for twice air exchange rate tests. The data show that the ACH in the test environment is about 0.55 (1/hr).

3.2 Airborne Bacterial Aerosol Concentration in the Testing Room

In order to carry out the environment antibacterial test of bacterial bioaerosols in the dentist dentist clinic operating environment, this work investigated the bacterial

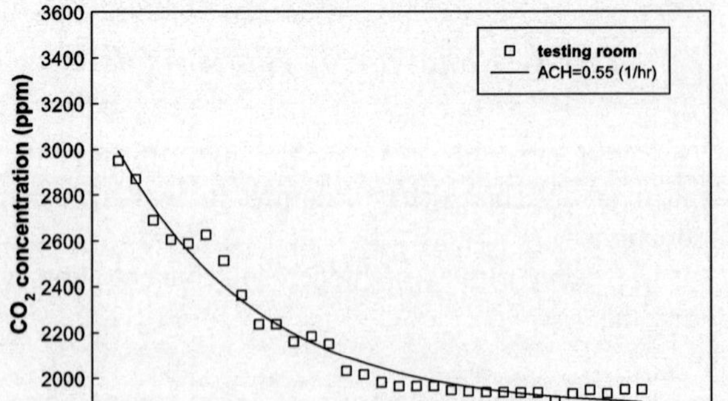

Fig. 1 Air exchange rate of the testing room

bioaeorsol distribution in the testing environment. This pre-experiment is a field sampling in the dentist clinic for 3 consecutive days. The sampling room was chosen in the dental consulting room. The air conditioner in this operating environment is a central air conditioning system. The sampling site is selected as the central area of the operating environment, and the sampling height is set to 1.5 m. Figure 2 showed the result of bacterial bioaerosol concentration distribution in the dental consulting room for 3 consecutive days. The results showed that bacterial bioaerosol concentration in the test environment for three consecutive days is 815–1,518 CFU/m^3. There is no significant difference in the sampling results of bioaerosols at different times. The main reason is that the air conditioner in the test environment belongs to the central air-conditioning system. Although there are no patients undergoing dental treatment at certain times, the bioaerogel concentration in the environment does not decrease significantly.

3.3 *Inactivation Ability of Carbon Nanotube Discharge Plasma System Against Bioaerosols*

Figure 3 showed that decay curve of bacterial biaoerosols in the dentist clinic testing room by using the nanotube carbon discharge plasma with operating voltages of 6.0 kV. In this antibacterial experiment, the nanotube carbon discharge plasma equipment was placed at a height of 1.5 m and about 0.5 m from the medical chair and plasma equipment. The sampling site was set at a height of 1.5, and 0.5 m and 1.0 m from the medical chair and plasma equipment.

According to the results of Fig. 3, the initial bacterial bioaerosol concentration is about 1,500 CFU/m^3. When using the nanotube carbon discharge plasma system, the bacterial bioaerosol concentration would decrease 70% with 120 min system working. The overall space bioaerosol inactivating ability is about 70%. The decay coefficient (ka) is about 0.0241 (min^{-1}).

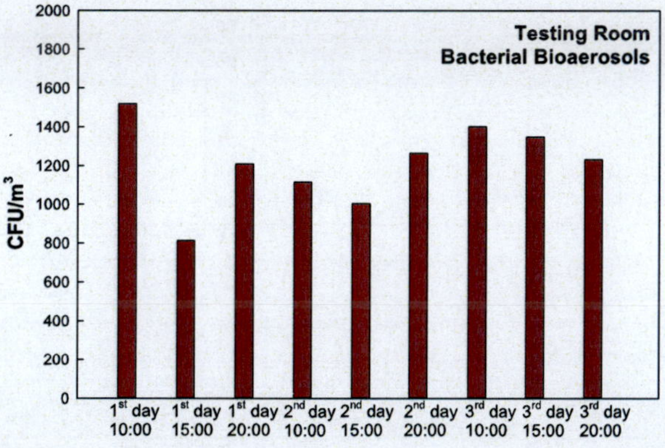

Fig. 2 Airborne bacterial aerosol concentration in the testing room

Fig. 3 Decay behavior of bacterial bioaerosols by applying the nanotube plasma in 1st test

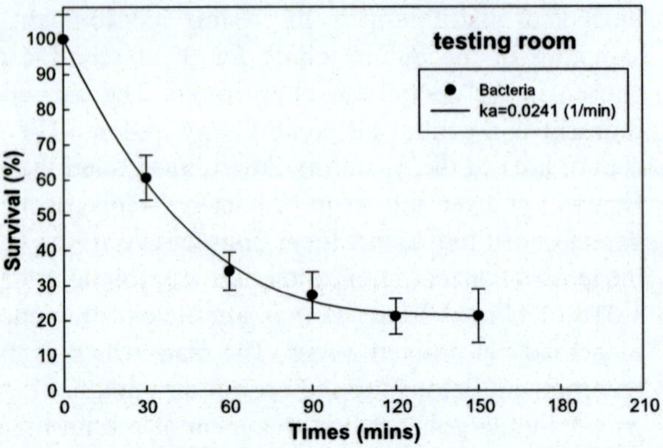

Figure 4 showed the 2nd test of the bacterial biaoerosols' decay behavior in the dentist clinic testing room by using the nanotube carbon discharge plasma with operating voltages of 6.0 kV. The initial concentration of indoor bacterial bioaerosols is in range of 1,000–1,500 CFU/m^3. In this 2nd test, the bioaerosol inactivating efficiency is 60% with 120 min system working. The decay coefficient (ka) is about 0.0195 (min^{-1}). According to the twice testing results, it indicated that this plasma system could inactivate bacterial bioaerosols effectively in 120 min. Furthermore, our pervious study showed that when indoor simulating chamber was using this system in 6.0 kV operating voltage, the decay constant (ka) for inactivating bacterial bioaerosols was 0.231 min^{-1} (Yang et al. 2011). The decay coefficient of the field test is only 10% of the simulating test. However, it has a significant inactivating effect on indoor bacterial bioaerosols.

Fig. 4 Decay behavior of bacterial bioaerosols by applying the nanotube plasma in 2nd test

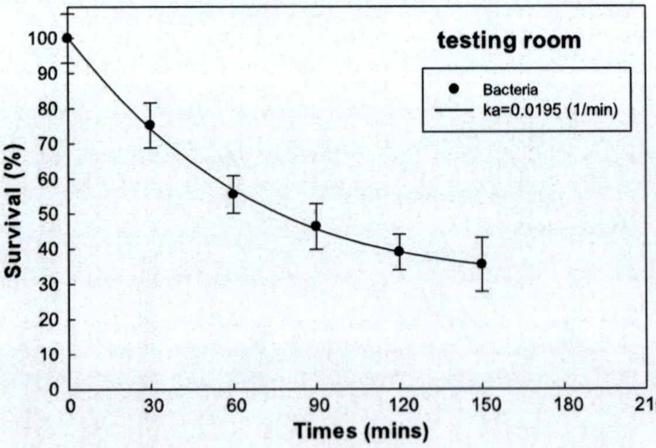

4 Summary

Within 120 min of applying nanotube plasma system, when the initial concentration of bacterial bioaerosol is 1,500 CFU/m^3, the concentration of bacterial bioaerosol will be reduced by about 70%. Results demonstrated that the nanotube discharge plasma system effectively inactivate bacterial bioaerosols in dentist clinic environments.

References

Timmerman, M., Menso, L., Steinfort, J., van Winkelhoff, A., van der Weijden, G.: Atmospheric contamination during ultrasonic scaling. J. Clin. Periodontol. 31, 458–462(2004).

Walker, J., Marsh, P. Microbial biofilm formation in DUWS and their control using disinfectants. J. Dent. 35,721–730(2007).

Rautemaa, R., Nordberg, A., Wuolijoki-Saaristo, K., Meurman, JH. Bacterial aerosols in dental practice e a potential hospital infection problem? J. Hosp. Infect. 64, 76–81(2006).

Chuang, CY., Cheng, HC., Yang, S., Fang, W., Hung, PC, Chuang SY. Investigation of the spreading characteristics of bacterial aerosol contamination during dental scaling treatment. J. Dent. Sci. 9(3), 294–296(2014).

Yang, S., Lee, WMG., Huang, HL., Huang, YC., Luo, CH., Wu, CC., Yu, K.P. Aerosol penetration properties of an electret filter with submicron aerosols with various operating factors. J. Environ. Sci. Health. A. 42(1), 51–57(2007).

Yang, S., Lee, WMG. Filtration characteristics of a fibrous filter pretreated with anionic surfactants for monodisperse solid aerosols. J. Aerosol Sci. 36(4), 419–437(2005).

Lin, CY., Li, CS. Control effectiveness of ultraviolet germicidal irradiation on bioaerosols. Aerosol Sci. Tech. 36, 474–478(2002).

Li, CS., Wang, YC. Surface germicidal effects of ozone for microorganisms. Am. Ind. Hyg. Assoc. J. 64, 533–537(2003).

K. Sunada, Y. Kikuchi, K. Hashimoto, A. Fujishima, Bactericidal and detoxification effects of TiO2 thin film photocatalysts, Environ. Sci. Technol. 1998,32, 726–728.

Lin, CY., Li, CS. Effectiveness of titanium dioxide photocatalyst filters for controlling bioaerosols. Aerosol Sci. Tech. 37(2), 162–170(2003).

Yu, KP., Lee, GWM., Huang, WM, Wu, CC, Lou, CL., Yang, S. Effectiveness of photocatalytic filter for removing volatile organic compounds in the heating, ventilation, and air conditioning system. J. Air Waste Manage. Assoc. 56(5), 666–674(2006).

Krueger, AP., Reed, EJ. Biological impact of small air ions. Science. 193(4259), 1209–1213 (1976).

Tanimura, Y. Inhibition of microbial growth using negative air ions. J. Antibact. Antifung. Agents. 25, 625–631(1997).

Chuang, CY., Yang, S., Huang, HC., Luo, CH., Fang, W., Hung, PC., Chung, PR. Applying the membrane-less electrolyzed water spraying for inactivating bioaerosols. Aerosol Air Qual. Res. 13(1), 350–359(2012).

Chuang, CY., Yang, S., Chang, MY., Huang, HC., Luo, CH., Hung, PC., Fang, W. Inactivation efficiency to Bacillus subtilis and Escherichia coli bacterial aerosols of spraying neutral electrolyzed water. J. Air Waste Manage. Assoc. 63(12), 1447–1456(2013).

Sato, M., Ohgiyama, T., Clements, JS. Formation of chemical species and their effects on microorganisms using a pulsed high-voltage discharge in water. IEEE Trans. Ind. Appl. 32(1), 106–112(1996).

Laroussi, M. Low temperature plasma-based sterilization: overview and state-of-the-art. Plasma Process Polym. 2(5), 391–400(2005).

Laroussi, M. Nonthermal decontamination of biological media by atmospheric-pressure plasmas: review, analysis, and prospects. IEEE Trans. Plasma Sci. 30(4), 1409–1415(2002).

Locke, BR., Sato, M., Sunka, P., Hoffmann, MR., Chang, JS. Electrohydraulic discharge and nonthermal plasma for water treatment. Ind. Eng. Chem. Res. 45(3), 882–905(2006).

Gadri, RB., Roth, JR., Montie, TC., Kelly-Wintenberg, K., Tsai, PPY., Helfritch, DJ., Feldman, P., Sherman, DM., Karakaya, F., Chen, Z. Sterilization and plasma processing of room temperature surfaces with a one atmosphere uniform glow discharge plasma (OAUGDP). Surf. Coat. Technol. 131(1-3), 528-541(2000).

Sladek, REJ., Stoffels, E. Deactivation of Escherichia coli by the plasma needle. J. Phys. D: Appl. Phys. 38(11), 1716(2005).

Gallagher, MJ., Vaze, N., Gangoli, S., Vasilets, VN., Gutsol, AF., Milovanova, TN., Anandan,S., Murasko, DM. Fridman, AA. Rapid inactivation of airborne bacteria using atmospheric pressure dielectric barrier grating discharge. IEEE Trans. Plasma Sci. 35(5), 1501–1510(2007).

Yang, S., Huang, YC., Luo, CH, Lin YC., Huang JW., Chuang, CJ,., Chuang CY. Inactivation efficiency of bioaerosols using carbon nanotube plasma. CLEAN 39(3), 201–205(2011).

Huang, YC., Luo, CH., Yang, S., Fang, W., Hung, PC. Removal of Bacillus Subtilis Bioaerosols by Using the Novel Plasma System. Adv. Sci. Lett. 19(9), 2741–2743(2013).

Gallagher MJ., Vaze, N., Gangoli, S., Vasilets, VN., Gutsol, AF., Milovanova, TN., Fridman, AA. Rapid inactivation of airborne bacteria using atmospheric pressure dielectric barrier grating discharge. IEEE Trans. Plasma. Sci. 35(5): 1501–1510(2007).

Spatial Distribution and Temporal Trends of Monthly Precipitation Concentration in Poyang Lake Basin

Liying Xiao, Guiqing Gao, Weilin Liu, and Yong Ji

Abstract The study on monthly distribution of annual precipitation is of great significance to drought or flood and water resources management. The 14-station monthly rainfall data in the Poyang Lake basin from 1957 to 2012 was collected. To analyze the distribution of monthly precipitation, the Precipitation Concentration Degree (PCD) and Precipitation Concentration Period (PCP) were adopted. Moreover, the modified Mann-Kendall (MMK) was applied to find the trend of precipitation concentration indices. It shows that these indices are applicable in this basin. The precipitation concentration detected by the PCD gradually increased from southwest to northeast in this basin. The variation range of precipitation concentration period is from early April to mid-June, with a downward trend from the west part to the east part. The precipitation concentration index shows an insignificant decrease trend in most of parts in this basin. Additionally, the result of the PCP shows an insignificant delay trend.

Keywords Precipitation concentration degree · Precipitation concentration period · Trend test · Poyang lake basin

1 Introduction

Global climate change can cause temporal and spatial changes of precipitation, which have been a lasting influence on water management, soil moisture and agriculture. Changes in the total amounts, patterns and intensity of precipitation can lead to abnormal weather (such as heavy rains, floods, and droughts). A small

L. Xiao (✉) · W. Liu · Y. Ji
School of Hydraulic and Ecological Engineering, Nanchang Institute of Technology, No. 289 Tianxiang Road, Nanchang, Jiangxi, China
e-mail: liying-xiao@163.com

G. Gao
School of Civil and Architecture Engineering, Nanchang Institute of Technology, No. 289 Tianxiang Road, Nanchang, Jiangxi, China

R. G. Gutierrez (ed.), *Wastewater Technologies and Environmental Treatment*, Springer Proceedings in Earth and Environmental Sciences,
https://doi.org/10.1007/978-3-030-61989-3_6

change in annual average precipitation may even induced to a large growth in the probability of extreme precipitation events. Moreover, extreme precipitation events may occur even though the annual average precipitation tends to decrease or stabilize under some special circumstances. Therefore, it is necessary to study the changes of precipitation patterns on regional and local scales.

Many papers have been carried out on the spatial and temporal uneven characteristics of precipitation. Scholars in various countries usually use the index of precipitation non-uniformity to evaluate the temporal and spatial variation of precipitation. Two precipitation concentration indices are often used to quantify the annual precipitation inhomogeneity, which are based on monthly scale or daily scale. Oliver (1980) proposed the monthly Precipitation Concentration Index (PCI) in 1980, which could be used to express the uniformity of the monthly rainfall within one year. In 2004, Martin Vide (2004) proposed the CI Index (a daily rainfall concentration index) to express the evenness Index of daily rainfall, indicating the contribution of the number of rainy days in a certain period (usually pointing to one year) to the total precipitation. PCI and CI are widely used in many countries. For example, Fatichi et al. (2002) used PCI and other indicators to study the relationship between the inter-annual changes of precipitation and the seasonal changes within the year. In 2015, Boucherf et al. (2014) analyzed the precipitation data in Algeria using CI indicators, and identified three climatic zones based on different CI ranges. In China, based on the principle of vector analysis, Zhang et al. (2003) developed the Precipitation Concentration Degree and Precipitation Concentration Period indices (the PCD and the PCP). Then, many researchers used these two indices to analyze the rainfall concentration degree and concentration period (Kong et al. 2015; Zhang et al. 2008, 2017). For example, Kong et al. (2015) analyzed the space-time pattern of concentration degree and period for the precipitation in China. Zhang et al. (2017) gained the spatiotemporal variation characteristics of concentration degree and period for the precipitation under complex terrains in Guizhou mountain area in 2017.

In recent years, extreme precipitation events occur frequently in the Poyang Lake basin, which has a great impact on the social and economic development of the basin. Many people have studied the variation characteristics of precipitation in the basin, but few papers focused on the spatial and temporal evolution of precipitation concentration closely related to flood, and drought disasters. Many researchers have put forward different precipitation concentration indicators from different perspectives, however, can these indicators be applied to the Poyang Lake basin? Based on this problem, this study will adopt the monthly PCD and PCP indices which are commonly applied in China to analyze the applicability and characteristics of these indices in the Poyang Lake Basin. At addition, an improved Mann-Kendall will be applied to analyze the variation trend of these indices. Results can provide key information for water resources management in the Poyang Lake basin.

2 Study Area

The Poyang Lake Basin is located in Jiangxi, China (115° 49′E to 116° 46′E and 28° 24′N to 29° 46′N). The drainage area of this basin is 16.22 × 104 km^2. It is located in the subtropical climate zone. The annual rainfall of this basin is about 1680 mm, and the spatiotemporal distribution of rainfall is uneven. According the annual rainfall, there are two obvious seasons—rainy season and dry season, and the transition period between the two seasons is relatively short. The monthly precipitation varies from 50.2 mm (in December) to 280.8 mm (in June), and the average monthly precipitation is 137.9 ± 74.2 mm. There was a significant transition in precipitation in June, and then a significant decrease. The precipitation from March to June accounted for about 54.2% of the total annual precipitation, while the precipitation from October to the next January only accounted for 15.9% of the total annual precipitation. Drought and flood are the most important extreme hydrological disasters in the basin. Droughts have been frequent in the basin in recent decades.

Monthly rainfall data from 14 stations during the period of 1957–2012 were collected (Fig. 1), and these monthly rainfall data are not missing. All these data were from the China Meteorological Administration.

3 Method

3.1 *Precipitation Concentration Degree and Period (The PCD and the PCP)*

The principle of monthly precipitation concentration degree and period (the PCD and the PCP) is to define the parameters of time allocation characteristics firstly,

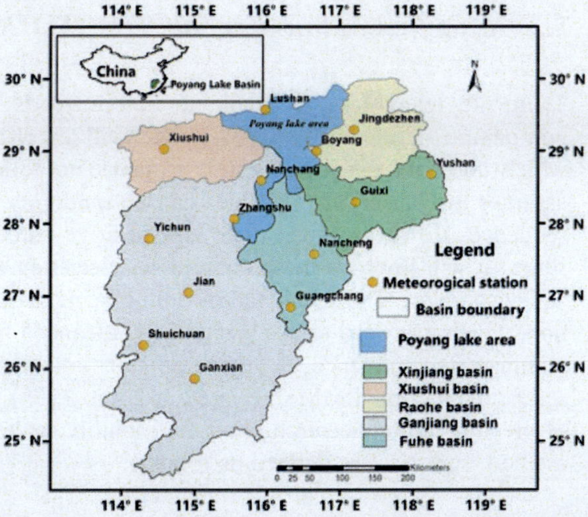

Fig. 1 Spatial distribution of meteorological stations

and then to quantitatively describe these two indices based on the vector principle. The concept of the PCD and the PCP refers to that the monthly precipitation is a vector, which has the length and the direction. The length refers to the total amount of precipitation per month, and the direction refers to the angle with an increment of 30° per month. A year's time forms a circle (i.e. 360°). The formula for calculating the PCD and PCP is as follows:

$$R_i = \sum r_{ij} \tag{1}$$

$$R_{xi} = \sum r_{ij} \sin \theta_j \tag{2}$$

$$R_{yi} = \sum r_{ij} \cos \theta_j \tag{3}$$

$$PCP_{ij} = \arctan\left(\frac{R_{xi}}{R_{yi}}\right) \tag{4}$$

$$PCD_{ij} = \frac{\sqrt{R_{xi}^2 + R_{yi}^2}}{R_i} \tag{5}$$

where, i refers to a certain year ($i = 1957, 1958, \ldots, 2012$) and j refers to a certain month ($j = 1, 2, \ldots, 12$). Here, r_{ij} means the precipitation amounts in the jth month and ith year and R_i means the precipitation amounts in the ith year (θ_j is the azimuth of the jth month). Thus, PCP_{ij} means the specific month for the precipitation concentration in the ith year, PCD_{ij} means the degree for the precipitation concentration in the ith year. The value range of PCD_{ij} is $0 \sim 1$. When $PCD_{ij} = 1$, the precipitation of a year is concentrated in a certain month. When $PCD_{ij} = 0$, it means that the annual precipitation is evenly distributed to each month.

3.2 *Modified Mann–Kendall Test (MMK)*

There are many tests available to detect trends of time series. One common non-parametric method is the Mann-Kendall test (Kendall et al. 1975; Mann 1945), which does not need the timing sequence to follow a specific distribution and assumes that all observed values can be randomly independent within the timing sequence. However, when there is a positive autocorrelation in the original time series, it will improve the probability of detecting a trend when there is no trend, and vice versa. The precipitation timing sequence often has obvious autocorrelation, which may lead to exaggerated significance of the trend of time sequence. To eliminate the influence of autocorrelation influences on the trend value of time series, the modified MMK (Hamed and Rao 1998) was adopted here to test trends of precipitation concentration on the monthly scale. In the original Mann-Kendall method, statistic S is defined as follows:

$$S = \sum_{k=1}^{n-1} \sum_{j=k+1}^{n} \operatorname{sgn}(x_j - x_k) \tag{6}$$

where x_k or x_j is the value of precipitation at time k or j (n is the length of time series) and $sgn(x)$ is defined as:

$$sgn(x) = \begin{cases} +1 & if\left(x_j - x_k\right) > 0 \\ 0 & if\left(x_j - x_k\right) = 0 \\ -1 & if\left(x_j - x_k\right) < 0 \end{cases} \tag{7}$$

Then, statistic S will follow a normal distribution when $n \geq 10$. Statistic Z_{MK} is defined as follows:

$$Z_{MK} = \begin{cases} \frac{S-1}{\sqrt{Var(S)}} & if \quad S > 0 \\ 0 & if \quad S = 0 \\ \frac{S+1}{\sqrt{Var(S)}} & if \quad S < 0 \end{cases} \tag{8}$$

In the MMK test, $Var(S)^*$ is modified as:

$$Var(S)^* = Var(S)\frac{n}{n^*} \tag{9}$$

$$\frac{n}{n^*} = 1 + \frac{2 * \sum_{i=1}^{n-1}(n-i)(n-i-1)(n-i-2)r_i}{n(n-1)(n-2)} \tag{10}$$

$$r_i = \frac{\frac{1}{n-i}\sum_{j=1}^{n-i}\left(x_j - \overline{x}\right)\left(t_{j+i} - \overline{x}\right)}{\frac{1}{n}\sum_{j=1}^{n}\left(x_j - \overline{x}\right)^2} \tag{11}$$

Here, r_i is the lag-i significant autocorrelation coefficient of the ith rank of actual time sequence (n^* is the size of time sequence). Then, the Z_{MMK} will be computed by Var(S)*. The trend is rejected at the significance level α of 0.05 when $|Z_{MMK}| > Z_{\alpha/2}$.

4 Results

4.1 Characteristics of the PCD and PCP Index

In order to understand the basic situation of the annual precipitation amounts and precipitation concentration indices in the Poyang Lake Basin, the annual precipitation and precipitation concentration indices calculated from 14 meteorological stations are averaged in the basin, and then the sequence of average annual

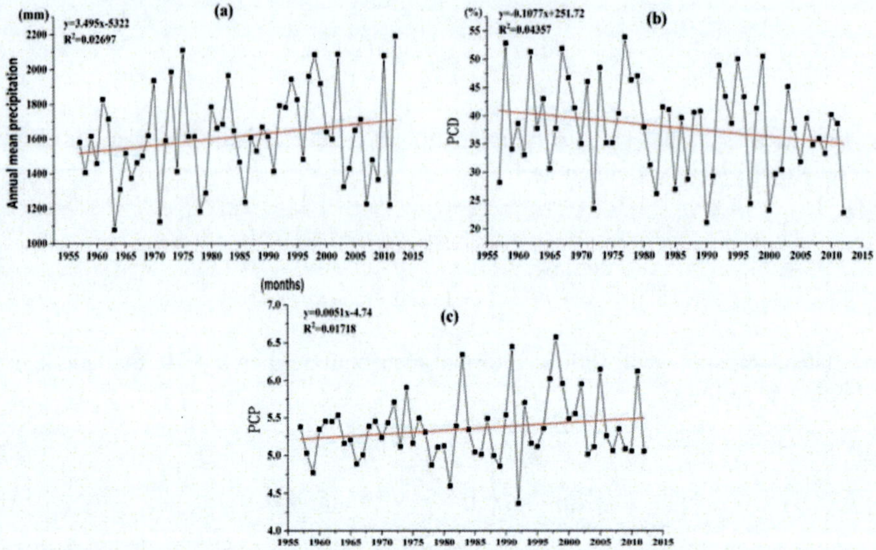

Fig. 2 Temporal changes of annual precipitation amounts and precipitation concentration indices from 1957 to 2012

precipitation and precipitation concentration indices in the basin can be obtained. Figure 2a gives the variation of annual precipitation sequence and there is an increasing trend in the annual precipitation from 1957 to 2012.

The inhomogeneity of annual rainfall amounts is analyzed with the rising trend in annual precipitation. Figure 2b and c show changes of the PCD and PCP index, which are commonly used in China. The fluctuation range of the PCD in the Poyang Lake basin is between 0.21 and 0.54, with an annual averaged value of 0.38. The minimum PCD value appeared in 1990 and the maximum in 1977, with an inter-annual variation coefficient of 0.22. The fluctuation range of the PCP is between 4.37(months) and 6.57(months), and the annual averaged value is 5.35 (months). The minimum PCP value appeared in 1992 and the maximum value appeared in 1998, with an inter-annual variation coefficient of 0.08. The linear trend shows that the precipitation concentration degree for the PCD decreased from 1957 to 2012. However, the precipitation concentration period for the PCP of the whole basin showed an upward trend, with an increasing rate of 0.0051/a, indicating that the future precipitation concentration period would be delayed.

4.2 *Spatial Distribution of the Annual Averaged PCD and PCP Indices*

The previous part of this paper analyzed the overall situation of the whole basin, and the following part will analyze the spatial differences of precipitation concentration degree within the basin. Figure 3 shows the spatial distribution of annual mean precipitation concentration indices. In Fig. 3a, the PCD index for the precipitation concentration varies from 0.33 to 0.41 over the years, with the maximum value at Boyang, Jingdezhen and Guangchang stations and the minimum value at Yichun and Suichuan stations. The PCD index of the whole basin presents a spatial variation trend of increasing from southwest to northeast.

As shown in Fig. 3b, the variation range of annual averaged value of PCP index for the precipitation concentration period is 5.05(month) to 6.11(month), with the maximum value at Lushan station and the minimum value at Nanchang, Zhangshu and Yushan stations. The PCP index of the whole basin shows a decreasing spatial variation trend from the west to the east. The precipitation of Lushan station and Suichuan Station concentrates in early June, while other stations mainly concentrate in May.

4.3 *Spatial Differences of Trends for the PCD and PCP Indices*

To detect the spatial differences of trends of rainfall concentration indices, the MMK method was used to test the precipitation concentration indices of 14 meteorological stations (significance level of 5%). The spatial distribution of trends for the precipitation concentration indices was also calculated (Fig. 4). As shown in Fig. 4a, from 1957 to 2012, a downward trend was found by using the PCD index for the precipitation concentration at all stations in the basin, and there is a significant decreasing trend at the Lushan station. As shown in Fig. 4b, an insignificant

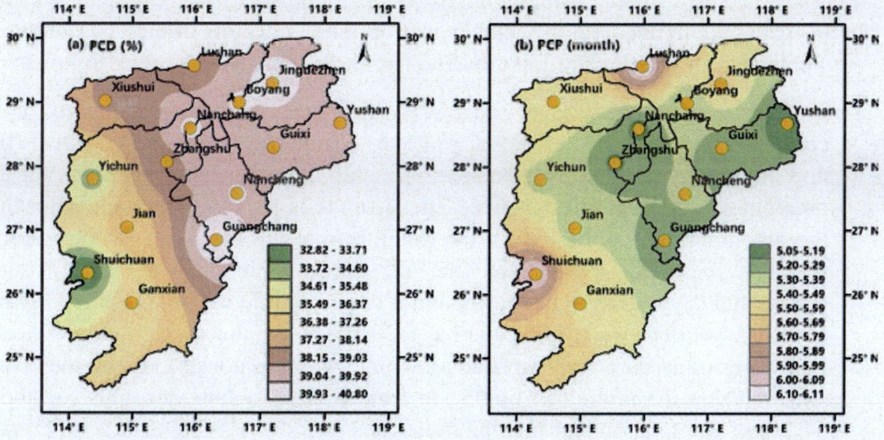

Fig. 3 Distribution of the annual mean PCD and PCP indices

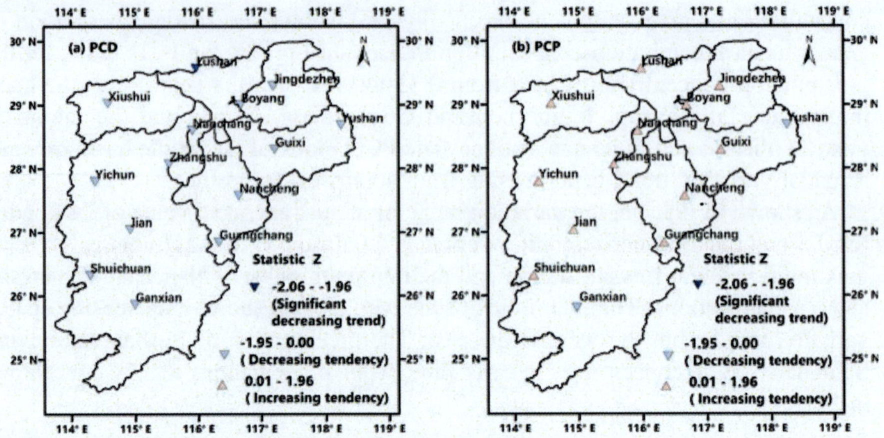

Fig. 4 Trends of the PCD and PCP indices

upward trend was also found by using the PCP index for the precipitation concentration period at most stations in the basin, and we can also find that there is an insignificant downward trend at Yushan station and Ganxian station.

5 Conclusions

The spatiotemporal variation characteristics of precipitation concentration in the Poyang Lake basin from 1957 to 2012 were studied by using the precipitation concentration indices in this paper. Results are basically consistent with those of other scholars in China (Yawen et al. 2014). At the same time, the research results of this paper are in line with the actual situation. The indicators of the PCD and PCP are applicable to the Poyang Lake basin. The main conclusions are as follows:

1. The temporal distribution of the PCD which describes the precipitation concentration degree in the Poyang Lake basin can objectively reveal the non-uniformity of monthly precipitation within the year. The PCD index for the precipitation concentration degree varies from 0.21 to 0.54, which indicates that the precipitation distribution on the monthly scale in the basin is not uniform within a year. This result is in line with the obvious characteristics of rainy season and dry season in the basin, and it is also in line with the results of Duan et al. (2014). The variation range of PCP index for the precipitation concentration period is from early April to mid-June, which is consistent with the actual situation that the rainfall in this basin from March to June accounts for about 54.2% of the total annual precipitation.

2. The spatial distribution of the PCD index in the basin shows a trend of gradual increase from southwest to northeast, while the precipitation concentration period shows a trend of gradual decrease from west to east. On the one hand, the PCD index for the precipitation concentration degree at all stations in the basin showed a decreased trend, and a significant decreased trend at Lushan station was found. On the other hand, the PCP index for the precipitation concentration period at most stations in the basin shows an insignificant upward trend. It indicates that the precipitation concentration period had a delayed trend.

Acknowledgements This research was supported by the Science and Technology Research Project of JiangXi Provincial Education Department (Program No. GJJ180924), and by the Teaching Reform Project of Nanchang Institute of Technology (Program No. 01001918), and by the Jiangxi Provincial Advantage Science and Technology Innovation Team Construction Project (Program No. 20171BCB24012), and by the National Natural Science Foundation of China (Program No. 51879128).

References

Oliver, J. E.: Monthly precipitation distribution: a comparative index. Professional Geographer 32(3), 300–309 (1980).

Martin-Vide, J.: Spatial distribution of a daily precipitation concentration index in peninsular Spain. International Journal of Climatology 24(8), 959–971 (2004).

Fatichi, S., Ivanov, V. Y., Caporali, E.: Investigating inter-annual variability of precipitation at the global scale: Is there a connection with seasonality. Journal of Climate 25(16), 5512–5523 (2002).

Boucherf, D., Benhamrouche, A., Hamadache, R., et al.: Spatial distribution of the daily precipitation concentration index in Algeria. Natural Hazards & Earth System Sciences Discussions 2(4), 2709–2729 (2014).

Zhang, L.J., Qian, Y.F.: Annual distribution features of the yearly precipitation in China and their interannual variations. Acta Meteorologica Sinica 17(2), 146–163 (2003).

Kong, F., Fang, J.Y., Liu, F., et al.: Variation in the spatiotemporal patterns of precipitation concentration degree and precipitation concentration period from 1951 to 2012 in china. Journal of Beijing Normal University (Natural science) 51(4), 404–411 (2015).

Zhang, B., Gu, X.P., Gu, S.H.: Spatial and Temporal Variation Characteristics of Precipitation Concentration Degree and Precipitation Concentration Period in Guizhou Mountain Area. Journal of China hydrology 6(37), 63–67 (2017).

Zhang, W., Zhang, T.Y., Liu, J.: Temporal and spatial features of precipitation concentration degree and precipitation concentration period of annual rainfall over northeast China. Journal of Nanjing Institute of Meteorology 31(3), 403–410 (2008).

Kendall, M.G.: Rank correlation methods. London, Griffin (1975).

Mann, H.B.: Nonparametric tests against trend. Econometrica 13(3), 245–259 (1945).

Hamed, K. H., Rao, A. R.: A modified Mann-Kendall trend test for autocorrelated data. Journal of Hydrology 204(1–4), 182–196 (1998).

Duan, Y.W., Zhu, K.Y., Ma, Z.G., et al.: Characteristics of Precipitation Concentration Index (PCI) Variations and Monthly Distribution of Annual Precipitation in China. Chinese Journal of Atmospheric Sciences 38 (6), 1124–1136 (2014).

Preliminary Research on Hydrogen and Oxygen Isotope Characteristics of River Waters in the Source Region of the Yangtze River and the Lancang River

Liangyuan Zhao, Wei Deng, Min Liu, Yuan Hu, and Lingxian Xie

Abstract By surveying and evaluating the δD and δ^{18}O values of the water in the source region of the Yangtze River and the Lancang River, the characteristics of the hydrogen and oxygen stable isotope composition have been analyzed. The results showed that the ranges of δD and δ^{18}O values varied from −126.28‰ to −47.96‰ and −16.77‰ to −7.13‰. The standard deviation of δD and δ^{18}O value are 18.58‰ and 2.34‰. The stability of δ^{18}O value is better than δD. Atmospheric precipitation is the principal source of river water supply in the region. The Local Meteoric Water Line equation is $\delta D = 7.73\ \delta^{18}O + 4.81$ ($R^2 = 0.96$, n = 19). Due to strong evapotranspiration, its slope and intercept are smaller than the Global Meteoric Water Line. The strong evapotranspiration in the source regions causes most of the *d-excess* to be negative, and the ice-snow melt water replenishment feature causes some *d-excess* to be positive. δD and δ^{18}O values in the source regions exhibited certain latitude effect and continental effect.

Keywords The source region of the Yangtze River · Hydrogen and oxygen isotopes · Deuterium excess parameter

1 Introduction

Water from different sources have different hydrogen and oxygen stable isotope characteristics. Owing to the isotope fractionation effect, the stable isotope undergo equilibrium fractionation and dynamic fractionation during the phase transition process, so the stable isotope of hydrogen and oxygen (D and ^{18}O) in the water is known as the demonstration of nature trace agent (Liu et al. 2008; Tian et al. 2008). In general, the composition of the D and ^{18}O in river water is constrained by the

L. Zhao (✉) · W. Deng · M. Liu · Y. Hu · L. Xie
Basin Water Environmental Research Department, Changjiang River Scientific Research Institute, Wuhan 430010, China
e-mail: zhaoliangyuannew@163.com

R. G. Gutierrez (ed.), *Wastewater Technologies and Environmental Treatment*, Springer Proceedings in Earth and Environmental Sciences,
https://doi.org/10.1007/978-3-030-61989-3_7

topography, landform and human activities, and its time-history changes and spatial distribution are different. At the same time, the characteristics of the D and ^{18}O of different supply sources such as precipitation, ice and snow melt water, groundwater, and the evapotranspiration effects of river water also affect the value of the D and ^{18}O in river water (Gu et al. 2011; Li et al. 2017).

The source region of the Yangtze River (SRYR) and the Lancang River (SRLR) are located in Qinghai-Tibetan Plateau, with an average altitude of more than 4000m. The SRYR, which is mainly composed of the Dangqu, Tuotuo and Chumar Rivers, is one of the most important area in the Three-river Headwater Region (TRHR) (Jiang et al. 2015; Chen 2002). The SRYR plays a significant role in maintaining the ecological balance of the whole Yangtze River Basin and promoting the sustainable use of water resources in the middle and lower reaches of the Yangtze River Basin (Zhao et al. 2019). In the present research, field investigations were carried out in 2019 for revealing the characteristics of the hydrogen and oxygen stable isotope composition in the rivers of SRYR and SRLR, aiming to provide references for scientific understanding of hydrologic processes in this area.

2 Sample Collection and Analysis

Field investigations in the SRYR and the SRLR were carried out in August, 2019. As shown in Fig. 1, 19 sampling sites were investigated in the SRYR and the SRLR. The geographic coordinate and altitude of sampling sites is displayed in Table 1.

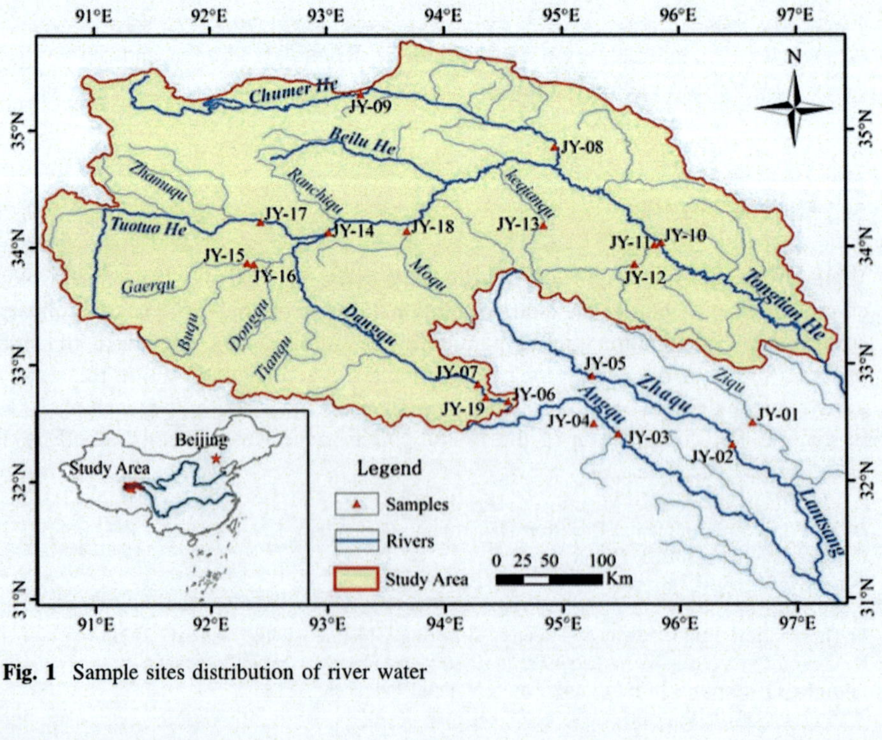

Fig. 1 Sample sites distribution of river water

Table 1 The geographic coordinate and altitude of sampling sites

Sampling site				Geographic coordinate		Altitude (m)
River name		Site number	River section	Latitude	Longitude	
The source region of the Lancang River (SRLR)		JY-01	Ziqu	32° 30′ 08″	96° 37′ 31″	3786
		JY-02	Nangqian	32° 18′ 43″	96° 27′ 18″	3609
		JY-03	Shamuyong	32° 24′ 19″	95° 28′ 39″	4008
		JY-04	Banayong	32° 29′ 34″	95° 16′ 05″	4225
		JY-05	Zaduo	32° 54′ 07″	95° 15′ 17″	4041
The source region of the Yangtze River (SRYR)	Dangqu	JY-06	The source of Dangqu	32° 41′ 08″	94° 32′ 42″	4681
		JY-07	The bridge of Dangqu	32° 53′ 03″	93° 50′ 52″	4722
		JY-15	Gaerqu	33° 51′ 57″	92° 21′ 01″	4530
		JY-16	Buqu	33° 51′ 55″	92° 22′ 01″	4530
		JY-19	Bama village	32° 52′ 08″	94° 12′ 55″	4807
	Chumar river	JY-08	The source of Chumar river	34° 51′ 18″	94° 56′ 30″	4217
		JY-09	Wudaoliang	35° 18′ 23″	93° 18′ 15″	4481
	Tuotuo river	JY-17	Tanggula Shan	34° 13′ 15″	92° 26′ 37″	4665
	Tongtian river	JY-10	Tongtian river	34° 01′ 25″	95° 49′ 20″	4054
		JY-11	Nieqiaqu	34° 01′ 25″	95° 49′ 18″	4004
		JY-12	Nieqiaqu (Ziduo county)	33° 51′ 10″	95° 37′ 17″	4136
		JY-13	Keqianqu	34° 11′ 10″	94° 50′ 39″	4474
		JY-14	Nangjibalong	34° 07′ 43″	93° 00′ 59″	4424
		JY-18	Moqu	34° 08′ 20″	93° 41′ 16″	4362

Three replicates of river water samples in each sampling point were collected and stored in dry and clean polyethylene bottles for laboratory analysis. The samples were measured with elemental analyzer (Flash EA2000) coupled with stable isotope mass spectrometer (Thermo DELTA V Advantage). Before the measurement, use 0.45 μm filter membrane to filter the samples. The measured δD accuracy is 0.4‰, and the $\delta^{18}O$ accuracy is 0.15‰. The ratio of the D and ^{18}O is expressed in terms of thousandths relative to Vienna's "Standard Mean Ocean Water" (VSMOW):

$$\delta X = \left(R_{sample} - R_{VSMOW}\right)/R_{VSMOW}$$

X refers to D or ^{18}O, R = $^{18}O/^{16}O$ or $^{2}H/^{1}H$, RVSMOW is the ratio of $^{18}O/^{16}O$ or $^{2}H/^{1}H$ in V_{SMOW}.

3 Results and Discussion

3.1 *Hydrogen and Oxygen Isotope Distribution Characteristics*

Generally, δD and δ^{18}O values are negative in most areas. As the distance of the water vapor source increases, the value will become more negative (Wen 2017). As shown in Fig. 2, the ranges of δD and δ^{18}O values varied from −126.28‰ to −47.96‰ and −16.77‰ to −7.13‰, which is consistent with the range given in the literature (Li et al. 2017). The standard deviation of δD is 18.58‰ and δ^{18}O value is 2.34‰, which indicates that the stability of δ^{18}O value is better than δD. The average value of δD is −59.69‰ and δ^{18}O is −12.20‰. More details are in Table 2.

3.2 *Characteristics of Precipitation Line*

The global atmospheric precipitation line refers to the linear correlation between the hydrogen-oxygen isotope composition of global atmospheric precipitation. In 1961, Craig (1961) conducted statistics and processing on sample data in different regions of the world, and obtained the Global Meteoric Water Line (GMWL) equation:

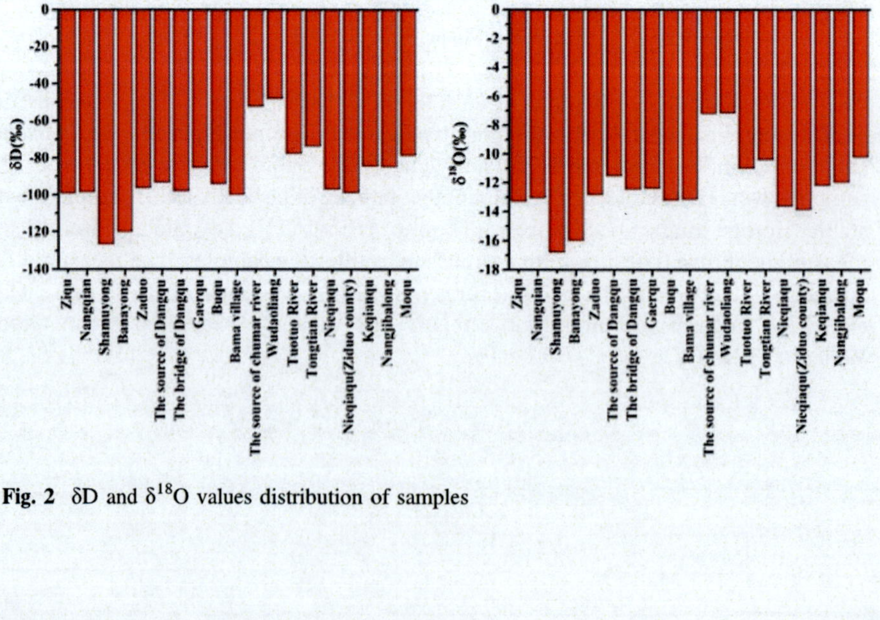

Fig. 2 δD and δ^{18}O values distribution of samples

Table 2 δD and $\delta^{18}O$ of different source regions

	δD		$\delta^{18}O$	
	Mean ± SD	Range	Mean ± SD	Range
The SRLR	−107.98 ± 12.56‰	−126.28‰ ~ −96.18‰	−14.37 ± 1.65‰	−16.77‰ ~ −12.81‰
The SRYR	−83.17 ± 15.82‰	−99.81‰ ~ −47.96‰	−11.43 ± 2.05‰	−7.13‰ ~ −13.84‰

$\delta D = 8\delta^{18}O + 10$. In 1993, Rozanski (1993) improved the GMWL equation to $\delta D = 8.17\ (\pm\ 0.17)\ \delta^{18}O + 11.27\ (\pm\ 0.65)$. Because different regions have different climatic conditions, such as water vapor sources, secondary evapotranspiration during precipitation and seasonal changes in precipitation, the atmospheric precipitation equations in different regions are quite different (Clark and Peter 1997). Wang (2000) obtained the equation of the meteoric water line in the middle and downstream of the Yarlung Zangbo River as $\delta D = 7.54\ \delta^{18}O + 15.92$. Wang (2019) studied the Dongkemadi River Basin in the SRYR and found that the Local Meteoric Water Line(LMWL) equation is $\delta D = 8.43\ \delta^{18}O + 20.62$.

As shown in Fig. 3, the hydrogen and oxygen isotopic composition of samples are distributed near the GMWL, indicating that atmospheric precipitation is the principal source of river water supply in the area. There is a significant linear relationship between δD and $\delta^{18}O$. Linear fitting can get the LMWL equation: $\delta D = 7.73\ \delta^{18}O + 4.81(R^2 = 0.96,\ n = 19)$, compared with the GMWL, its slope and intercept are smaller. The δD and $\delta^{18}O$ values of some samples are located at the lower right of the GMWL. Because the river is located in a semi-arid environment, it is subject to strong evapotranspiration, which increases the content of D and ^{18}O isotopes (Cappa et al. 2003).

3.3 Deuterium Excess Parameter Characteristics

The slope of the atmospheric precipitation line reflects the difference between the different fractionation degrees of δD and $\delta^{18}O$, and the intercept indicates the degree to which the hydrogen and oxygen isotopes deviate from equilibrium due to dynamic fractionation (Zhang et al. 2005). In 1964, Dansgaard (1964) first proposed the concept of deuterium excess, defining the deuterium excess parameter(*d-excess*) as $d = \delta D - 8\ \delta^{18}O$, that meant the *d-excess* is equivalent to the intercept when the slope of the atmospheric precipitation line in a certain area is 8. Different

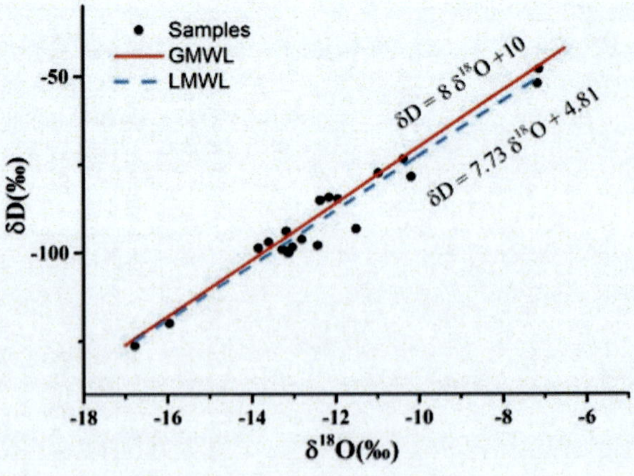

Fig. 3 Relationship diagram of hydrogen and oxygen isotopes

regions have different *d-excess*. Therefore, the *d-excess* can intuitively reflect the degree of imbalance in the process of evapotranspiration and condensation of regional atmospheric precipitation.

As shown in Fig. 4, The *d-excess* of most samples are less than the global average (d = 10‰). The sample collection time is during the melting period (May–September), when the amount of evaporation is large (Zhang 2004). It may be the river is located at the high-altitude area and shows strong evapotranspiration, resulting in smaller *d-excess*. The *d-excess* of some water samples are close to or exceeding 13‰, which may be due to the melting of some glaciers and the increase of the proportion of ice and snow melt water supply rivers, resulting in a partial *d-excess* being positive (Yu et al. 2010).

There are many plateau lakes in the upper Chumar River. The average annual precipitation of Chumar River is about 257 mm, and the average annual evapotranspiration is 1550 mm (Li 2013). Because of strong evapotranspiration, the *d-excess* of the upstream sample point (JY-8) is significantly less than 10‰. The Tuotuo River mainly receives glacial melt water. The average annual precipitation is about 282.3 mm, and the annual average annual evapotranspiration is 1646.1 mm (Wu 2012). The *d-excess* is slightly greater than 10‰ due to the replenishment of glacier melt water. Dangqu has abundant groundwater recharge and precipitation recharge. The average precipitation for many years is 400 mm (Shen et al. 2009; Chen 2014). Due to the precipitation process is affected by secondary evapotranspiration, and the *d-excess* of the samples (JY-6 and JY-7) collected at the source are obviously negative. In addition, the soil moisture at the location of the groundwater supply is relatively large, which makes the *d-excess* more negative (Zhang et al. 2019).

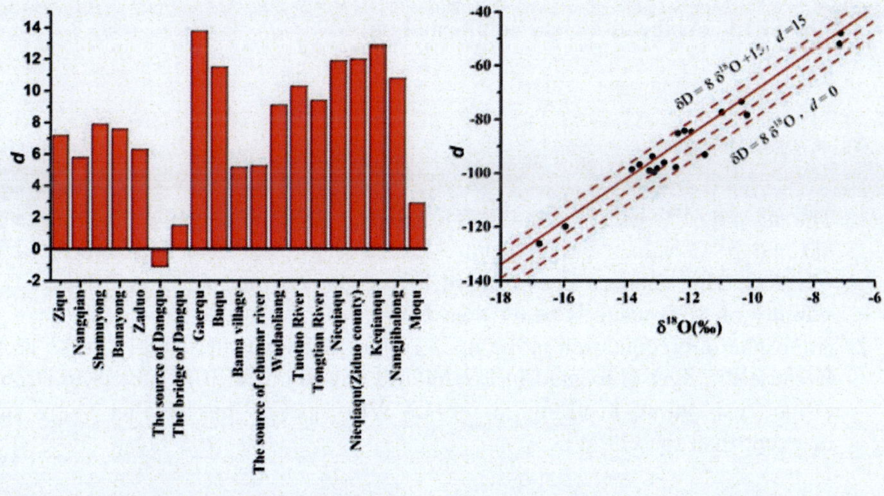

Fig. 4 Distribution of *d-excess* in samples

Table 3 δD and $\delta^{18}O$ correlation analysis with altitude, latitude and longitude

	Altitude	Latitude	Longitude
δD	$R^2 = 0.0532$	$R^2 = 0.7167$	$R^2 = 0.1325$
$\delta^{18}O$	$R^2 = 0.0652$	$R^2 = 0.5605$	$R^2 = 0.1015$

3.4 Elevation Effect, Latitude Effect and Continental Effect

The results of studies on stable isotopes of atmospheric precipitation indicate that there are elevation effects, latitude effects and continental effects of stable isotopes in precipitation. The phenomenon that $\delta^{18}O$ gradually decreases with increasing altitude in atmospheric precipitation is called the elevation effect, that is, as the altitude increases, heavy isotopes content in precipitation gradually decreases. With the increase of latitude, the content of heavy isotopes in precipitation is getting less and less, which is called the latitude effect. Table 3 shows the correlation analysis of the sample's hydrogen and oxygen isotope ratios with altitude, latitude and longitude.

It can be seen from Table 3 that δD or $\delta^{18}O$ has no obvious correlation with altitude or longitude, but has a positive correlation with latitude. The linear regression equation of δD and latitude is $y = 18.02x - 694.41$ ($R^2 = 0.7167$). The linear regression equation of $\delta^{18}O$ and latitude is $y = 2.01x - 79.61$ ($R^2 = 0.5605$). This indicated that the δD or $\delta^{18}O$ values exhibited a certain latitude effect.

Under the large hydrological cycle scale, different regions exhibit different δD and $\delta^{18}O$ values. Generally, the water samples farther away from the water vapor source have a smaller value. The study areas is located in the inland high altitude area. According to the analysis of the distribution characteristics, δD and $\delta^{18}O$ are significantly negative (the ranges of δD and $\delta^{18}O$ values varied from −126.28‰ to −47.96‰ and −16.77‰ to −7.13‰). This showed that the river water in the SRYR and the SRLR exhibited a clear continental effect.

4 Conclusion

1. The δD and $\delta^{18}O$ values in the SRYR and the SRLR are negative. The ranges of δD and $\delta^{18}O$ values varied from −126.28‰ to −47.96‰ and −16.77‰ to −7.13‰. The standard deviation of δD is 18.58‰ and $\delta^{18}O$ is 2.34‰. The stability of $\delta^{18}O$ value is better than δD.
2. Atmospheric precipitation is the main source of river water supply in the area. Because the river is located in a semi-arid environment, it is subject to strong evapotranspiration, resulting in the LMWL equation has smaller slope and intercept than the GMWL.

3. The strong evapotranspiration in the source regions causes most of the *d-excess* to be negative, and the ice-snow melt water replenishment feature causes some *d-excess* to be positive. Due to the difference in replenishment methods, precipitation and evapotranspiration, the *d-excess* changes significantly in the SRYR.
4. The δD and $\delta^{18}O$ values in the source regions exhibited certain latitude effect and continental effect, and the δD and $\delta^{18}O$ values have no obvious correlation with altitude or longitude.

Acknowledgements This work was financially supported by Central Public-interest Scientific Institution Basal Research Fund (Grant No. CKSF2019292/SH, CKSF2017062/SH), Outstanding Young Talents of National High-level personnel of special support program(CKSD2019542/SH).

References

Liu, Z., Tian, L., Yao, T.: The temporal and spatial variations of $\delta^{18}O$ in river water of the Yarlung Zangbo River Basin[J]. Journal of Glaciology and Geocryology 30(1), 20–27(2008).
Tian, L., Ma, L., Yu, W.: Seasonal Variation of Stable Isotopes and Water Vapor Transport in Yushu Precipitation in the Eastern Tibetan Plateau[J]. Science in China Series D: Earth Sciences (08), 986–992(2008).
Gu, W., Pang, Z., Wang, Q.: Isotope Hydrology[M]. Beijing: Science Press, 619–621(2011).
Li, J., Zhao, J., Yang, L.: Stable Hydrogen and Oxygen Isotopes in Waters in Tibet[J]. Arid Zone Research 34(02), 411–415(2017).
Jiang, L., Yao, Z., Liu, Z.: Hydrochemistry and its controlling factors of rivers in the source region of the Yangtze River on the Tibetan Plateau[J]. Journal of Geochemical Exploration 155, 76–83(2015).
Chen, X.: Eco-Environment of Nature Reserves in the Three Rivers' Source[M]. Qinghai People's Publishing House, Xining (2002).
Zhao, L., Li, W., Lin, L.: Field Investigation on River Hydrochemical Characteristics and Larval and Juvenile Fish in the Source Region of the Yangtze River[J]. Water 11(7), 1342(2019).
Wen, X.: Study on Thermodynamics and Stable Isotopes of Water Body in the Lugu Lake[D]. Yunnan Normal University, (2017).
Craig, H.: Isotopic variations in meteoric waters[J]. Science, 133(3465):1702–1703(1961).
Rozanski, K., Araguás- Araguás, L., Gonfiantini, R.: Isotopic patterns in modern global precipitation[J]. Climate change in continental isotopic records 78, 1–36(1993).
Clark, D., Peter, F.: Environmental isotopes in hydrogeology[M]. CRC press, New York, (1997).
Wang, J., Liu, T., Yin, G.: Characteristics of isotope distribution in precipitation in the middle-lower reaches of Yarlung Zangbo River[J]. Earth and Environment (01), 63–67(2000).
Wang, S., Wang, Q., Wu, J.: Characteristics of Stable Isotopes in Precipitation and Moisture Sources in the Headwaters of the Yangtze River[J]. Environmental Science 40(06), 2615–2623 (2019).
Cappa, C., Hendricks, M., DePaolo, D.: Isotopic fractionation of water during evapotranspiration [J]. Journal of Geophysical Research: Atmospheres 108(16), (2003).
Zhang, X., Sun, W., Liu, J.: Stable isotopes in precipitation in the vapor transport path in Kunming of southwest China[J]. Resources and Environment in the Yangtze Basin 14(5), 665–669 (2005).
Dansgaard, W.: Stable isotopes in precipitation[J]. Tellus 16(4), 436–468(1964).

Zhang, H.: Research of evapotranspiration characteristics and its variation tendency in the source of the Yangtze river and the region around[J]. Journal of Water Resources and Water Engineering 24(3), 49–52(2004).

Yu, T., Gan, Y., Zhou, A.: Characteristics of oxygen and hydrogen isotope distribution of surface runoff in the Lhasa River Basin[J]. Earth Science 35(5), (2010).

Li, Z.: Fluvial Processes and wetland Degradation Mechanism of the Sanjiangyuan Source[D]. Tsinghua University, (2013).

Wu, S.: Response of glacier changes to climate change and their influence on the runoff in Tuotuo River Basin[D]. University of Chinese Academy of Sciences, (2012).

Shen, Y., Wang, G., Wang, G.: Impacts of climate change on glacial water resources and hydrological cycles in the Yangtze River source region, the Qinghai-Tibetan Plateau, China: A Progress Report[J]. Sciences in Cold and Arid Regions 1(6), 0475–0495(2009).

Chen, J.: One of the Source of Yangtze River: Dangqu River Basin and Its Ecological System[J]. Yangtze River Scientific Research Institute 31(10), 1–6(2014).

Zhang, H., Yu, K., Li, Z.: Characteristics of Hydrogen and Oxygen Isotopes in Different Water Bodies in Hilly and Gully Regions of the Loess Plateau[J]. Environmental Science 40(07), 3030–3038(2019).

Evaluating Remanufacturing Lithium-Ion Batteries

Meihan Yu, Bo Bai, and Xiaoming Ma

Abstract Waste management of lithium-ion batteries (LIBs) after retirement is worrying as the wide application of electric vehicles. More and more attention has been paid to the remanufacturing of LIBs, because it can reduce the environmental burden of spent LIBs and the risk of raw materials supply when remanufacturing LIBs. This paper use the life cycle assessement to evaluate the environmental impacts and employ the process-based cost model to calculate the economic influence. Results show that remanufacturing NCM_{111} battery with pyrometallurgical recycling method could decrease the greenhouse gas (GHG) emission by 4.76% and reduce costs by 8.11%. In addition, this study uses sensitivity analysis method to analyze the influence of battery types. The cost of remanufacturing NCM_{811} is the least among three types of LIBs, which costs 19.72 dollars per cell. Remanufacturing NCM_{622} emits the most GHG among three types of LIBs, which emits 11,009.75 g GHG per cell. In a word, remanufacturing LIBs is environmentally-friendly and commercially-profitable.

Keywords Lithium-ion battery · Environmental impact · Economic benefit · Remanufacturing · Pyrometallurgical recycling method

M. Yu · B. Bai · X. Ma (✉)
School of Environment and Energy, Peking University Shenzhen Graduate School, Shenzhen, Guangdong 518055, China
e-mail: xmma@pku.edu.cn

B. Bai · X. Ma
College of Environmental Science and Engineering, Peking University, Yiheyuan Road, Beijing 100871, China

R. G. Gutierrez (ed.), *Wastewater Technologies and Environmental Treatment*, Springer Proceedings in Earth and Environmental Sciences,
https://doi.org/10.1007/978-3-030-61989-3_8

1 Introduction

The increasingly widespread use of electric vehicles (EV) as a sustainable means of transportation (Gu et al. 2017) has freed China's transportation sector from traditional fossil fuels, thereby reducing the burden of GHG emissions in China (Li et al. 2018; Qiao et al. 2019).

In China, production and sales of electric vehicles is growing rapidly, exceeding 1.256 billion in 2018 and 3.8 times more than 331 million in 2015. By 2020, the target is to sell 5 million electric vehicles (Council et al. 2012). Due to the unprecedented continuous expansion of electric vehicles and the limited service life of lithium-ion batteries (Saxena et al. 2015), the future disposal of large-scale batteries will cause people to worry about waste management of waste batteries.

Remanufacturing, i.e. the production of lithium-ion batteries from waste battery recycling materials, can reduce the burden of metal supply (Ziemann et al. 2018), environmental benefits and economic feasibility (Xiong et al. 2020). Remanufacturing generates recycled materials that can be re-used for manufacturing and therefore is able to reduce the demand for virgin materials (Ziemann et al. 2018), which may reduce the demand for imported materials (Habib et al. 2016; Olivetti et al. 2017).

While recognizing the importance of recycling subjectively, more quantitative analysis is needed to determine the feasibility of producing batteries with recycled materials instead of raw materials. Previous studies on the GHG emission of battery recycling are listed in Table 1, which indicate that the calculation deviation is large. From an economic point of view, the economic benefits of remanufacturing are still controversial. Xiong et al. (2020) calculated that the cost could reduce \$1.87 kg^{-1} battery produced when remanufacturing batteries. Zhu and Yu (2019) found that high-capacity recyclers, those who gain remanufactured raw materials by disposing of spent batteries, make a profit. Moreover, battery manufacturer could earn more profits with the increasing proportion of large-capacity recyclers. However, other pessimistic studies believe that the increasing proportion of the manufacturing with recycled material will reduce the revenue; therefore, the profit of remanufacturer is generally lower than that of manufacturer (Gu et al. 2018a, b; Song et al. 2017). The potential reasons that cause the large deviations of environmental and economic assessments are: (i) different recycling methods considered in those studies, (ii) battery chemistry and (iii) regional disparity.

In recent years, battery technology has developed rapidly, so the past research is inconsistent with the current situation, which needs to be updated according to the existing battery technology. Secondly, there are great differences in different regions, which need to be explored according to China's situation. Thirdly, different recycling technologies have a great impact on remanufacturing. The paper of Xiong et al. (2020) mainly explores the environmental and economic impacts of remanufacturing LIBs with hydrometallurgical recycling in China, but does not consider the pyrometallurgical recycling method which is also popular. This paper focuses on the environmental and economic impacts of remanufacturing LIBs with pyrometallurgical recycling method.

Table 1 Summary of the literature on environmental assessment of LIB recycling

Reference	Recycling processes	Battery technologies	Results
Qiao et al. (2019)	Hydrometallurgical recycling	NCM	The profit of recycling LIBs: 1203.1 (dollar per EV) The GHG emission reduction of recycling LIBs: 1109.7 (kg CO_2e per EV)
Dunn et al. (2012)	Hydrometallurgical, intermediate physical, and direct physical recycling	$LiMn_2O_4$ (LMO)	Direct physical recycling, which has the most environmental benefits among those three recycling methods, could decrease about 40% energy consumption
Ciez and Whitacre (2019)	Hydrometallurgical, process. pyrometallurgical, and direct physical recycling	NMC_{622}, NCA and LFP	Although pyrometallurgical and hydrometallurgical recycling processes cannot achieve significant reduction of GHG emissions, direct physical recycling could decrease GHG emission and cost
Dunn et al. (2015)	Pyrometallurgical, intermediate physical, and direct physical recycling	LMO, LFP, NCM, $LiCoO_2$ (LCO), and LMR-NMC cathodes	The GHG emission reduction of pyrometallurgical, intermediate physical, and direct physical recycling is 60–75%, 11–91% and 81–98%, respectively

2 Methods

2.1 *Life Cycle Assessment (LCA)*

LCA has been widely used to calculate the environmental impact of products and processes throughout their life cycle (Clift 2006). Based on GREET model (ANL 2018a) and BatPaC model (ANL 2018b), Everbatt model (ANL 2018c) could be used to analyze the environmental impact and economic influence of remanufacturing LIBs. In the life cycle of this study, remanufacturing LIBs with recycled materials includes three stages: (I) recycling; (II) cathode remanufacturing; (III) battery remanufacturing (Fig. 1). After recycling and cathode remanufacturing, recycled cathode materials are produced.

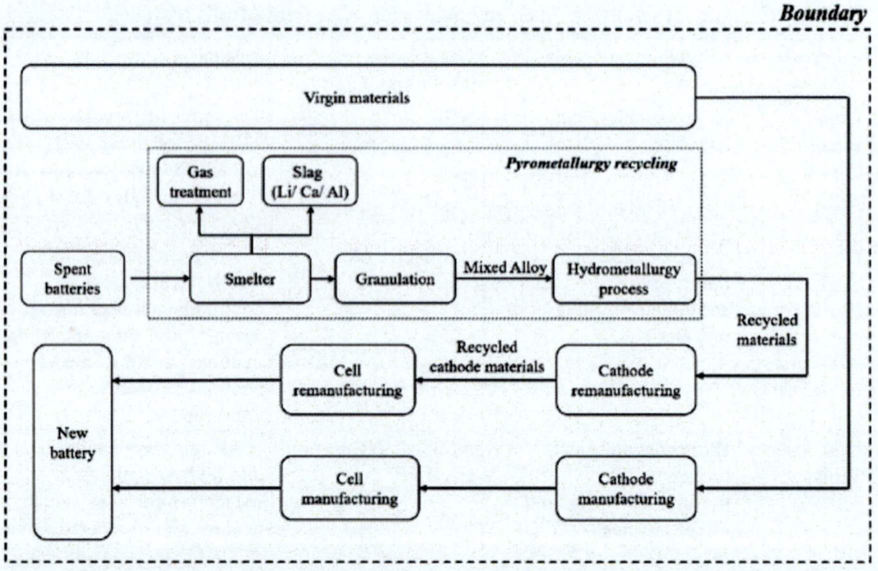

Fig. 1 Manufacturing and remanufacturing process of LIBs

2.2 Remanufacturing Process

In the recycling process, pyrometallurgical recycling method is used to recycle materials from spent LIBs. Pyrometallurgical recycling uses high temperature to promote the redox process. Coke, sand, slag, and limestone are burnt with the batteries to produce mixed metal alloys. Through physical and chemical reactions, some substances like cobalt and nickel are recycled. However, the aluminum and lithium will be eventually in slag because recovering it from slag is an economically infeasible process (Linda 2018). In addition, other substances, such as electrolyte, polyvinylidene fluoride and so on, are burned in furnaces to give energy for the reaction process.

In the cathode remanufacturing stage, the output of virgin materials and recycled materials is assumed to be 10,000 tons/year. Except for the cathode manufacturing process, the energy and material input of the remanufacturing LIBs with virgin materials are the same as that with recycled materials.

2.3 GHG Emissions

The study measured greenhouse gas emissions (g) per kilogram of LIBS over the life cycle. In order to calculate life cycle carbon, this study employ the default values of EverBatt model that are derived from previous studies or process analysis (Dunn et al. 2012, 2015).

As in Xiong et al. (2020), the environmental impact are calculated based on Eq. (1):

$$EF = \sum_i m_i * cw_i + \sum_j q_j * cw_j + P \quad (1)$$

where EF is the greenhouse gas emission; m_i is the mass of the material used; q_j is the amount of fuel j used; cw_i is the GHG emission coefficient (intensity) of material i and cw_j is the GHG emission coefficient (intensity) of fuel j; P is the greenhouse emission due to combustion or thermal decomposition of material.

2.4 Cost Model

Process-based cost model was used to evaluate the overall cost of remanufacturing LIBs. The total cost consists of five parts: (i) material and energy inputs, (ii) equipment and construction investment, (iii) labor costs, (iv) maintenance, and (v) fixed indirect costs (Ciez and Whitacre 2017). The material price is derived from the market information and default value of EverBatt model. The total cost of remanufacturing LIBS is calculated according to formula (2).

$$\mathrm{TC} = C_{c\&t} + C_r + C_{cathode} + C_{cell} \quad (2)$$

where TC refers to total cost of remanufacturing LIBs; $C_{c\&t}$ denotes collection and transportation cost; C_r represents recycling cost; $C_{cathode}$ refers to the cost of cathode remanufacturing; C_{cell} denotes the cost of cell remanufacturing.

3 Results

3.1 GHG Emission

Manufacturing LIBs with virgin material emit 11,138.74 g CO_2e per kg cell. The GHG emission of remanufacturing LIBs with pyrometallurgical recycling method is 10,608.66 g/kg, which is 4.76% lower than that of lithium battery made from virgin materials. The results indicate that compared with manufacturing LIBs with virgin materials, remanufacturing LIBs with pyrometallurgical recycling method is environmentally-friendly.

The main difference between remanufacturing LIBs and manufacturing batteries with virgin materials lies in the way in which raw materials are obtained. One is to use recycled materials, the other is to use virgin materials. Recycled materials can reduce GHG emissions because recycling process of it emits lower GHG emissions than mining. Therefore, if the carbon emission in the recycling process is greatly reduced, the environmental comparative advantage of manufacturing LIBs with recycled materials is likely to increase. Therefore, we should first identify the parts of the recycling process that have a great impact on the environment, so as to put forward specific carbon reduction plans for each part. Figure 2 shows the GHG emissions of each part in the recycling process.

As shown in Fig. 2, smelting process generate large amount of GHG emission, which mainly from combusting of materials. The use of hydrochloric acid is the second largest part of GHG emissions.

3.2 Cost

The cost of manufacturing LIBs is 23.83 dollars. Remanufacturing LIBs with pyrometallurgical recycling method cost 21.90 dollars, which reduce by 8.11%. The results indicate that remanufacturing LIBs with pyrometallurgical recycling method is economically viable.

3.3 Sensitivity Analysis

Figure 3 shows the GHG emission of remanufacturing NCM_{111}, NCM_{622} and NCM_{811} battery. The GHG emission of manufacturing LIBs with virgin materials is 11,138.74 g/kg, 11,308.30 g/kg and 11,186.41 g/kg. Therefore, manufacturing NCM_{622} battery emits most GHG. The GHG emission of remanufacturing NCM_{111}, NCM_{622} and NCM_{811} battery is 10,608.66 g/kg, 11,009.75 g/kg and 10,853.05 g/kg. Compared with manufacturing LIBs with virgin materials, remanufacturing

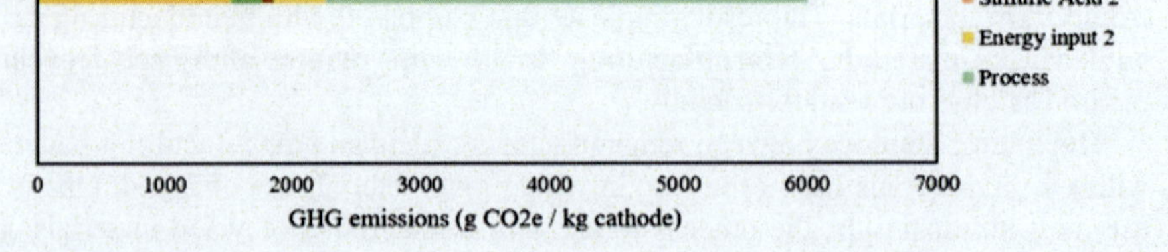

Fig. 2 Breakdowns of GHG emissions for recycling process. "Process" in the diagram means smelting process

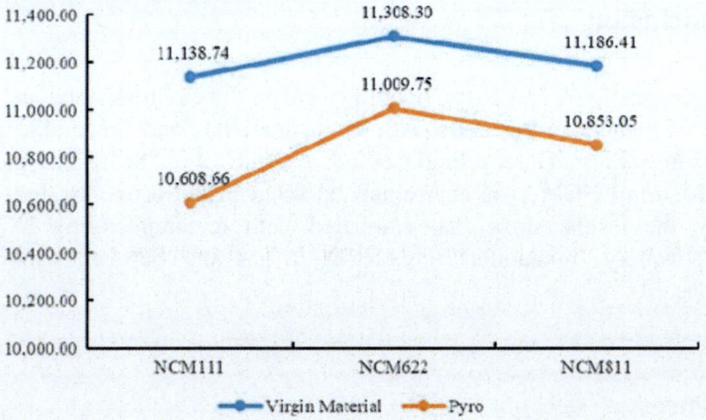

Fig. 3 GHG emission of remanufacturing NCM_{111}, NCM_{622} and NCM_{811} battery (unit: g/kg)

NCM_{111}, NCM_{622} and NCM_{811} battery could reduce by 4.76%, 2.64% and 2.98% GHG emission. The results indicates that remanufacturing NCM_{111} have the best emission reduction effect, followed by NCM_{811}.

Figure 4 shows the cost of remanufacturing NCM_{111}, NCM_{622} and NCM_{811} battery. The cost of manufacturing LIBs with virgin materials is $23.83, $21.94 and $20.50, which indicates that as the content of cobalt decreases, the cost of manufacturing NCM battery has dropped. Remanufacturing NCM_{111}, NCM_{622} and NCM_{811} battery with pyrometallurgical recycling method cost $21.90, $20.96 and $19.72. Compared with manufacturing LIBs with virgin materials, the cost of remanufacturing NCM_{111}, NCM_{622} and NCM_{811} battery could reduce by 8.11%, 4.47% and 3.77%. As a result, remanufacturing NCM_{111} have the best cost reduction effect, followed by NCM_{622}.

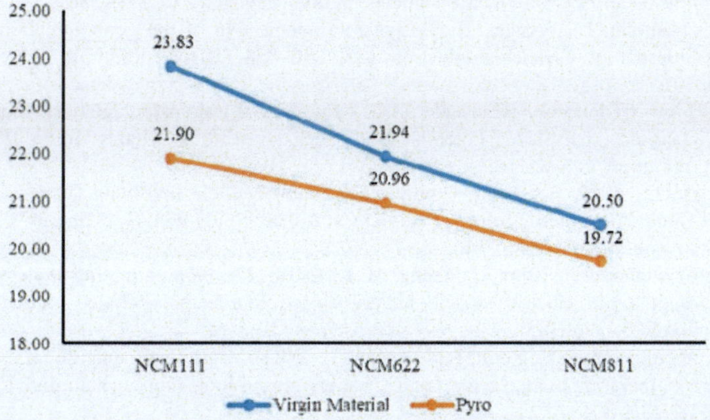

Fig. 4 Cost of remanufacturing NCM_{111}, NCM_{622} and NCM_{811} battery (unit: $)

4 Conclusion

This paper uses LCA and cost model to calculate the environmental and economic impacts of manufacturing LIBs with virgin materials and remanufacturing LIBs with pyrometallurgical recycling method, respectively. The results indicate that remanufacturing NCM_{111} is environmental-friendly and economically-feasible. In addition, the results show that compared with remanufacturing NCM_{622} and NCM_{811} battery, remanufacturing NCM_{111} has the best emission and cost reduction.

References

Gu, H., Liu, Z., Qing, Q.: Optimal electric vehicle production strategy under subsidy and battery recycling. Energy Policy 109, 579–589 (2017). doi:https://doi.org/10.1016/j.enpol.2017.07.043.

Li, L., Dababneh, F., Zhao, J.: Cost-effective supply chain for electric vehicle battery remanufacturing. Applied Energy 226, 277–286 (2018). doi:https://doi.org/10.1016/j.apenergy.2018.05.115.

Qiao, Q., Zhao, F., Liu, Z., Hao, H.: Electric vehicle recycling in China: Economic and environmental benefits. Resources, Conservation and Recycling 140, 45–53 (2019). doi:https://doi.org/10.1016/j.resconrec.2018.09.003.

Council, G.O.o.t.S.: Energy Saving and New Energy Vehicle Industry Planning from 2012 to 2020 (2012).

Saxena, S., Le Floch, C., MacDonald, J., Moura, S.: Quantifying EV battery end-of-life through analysis of travel needs with vehicle powertrain models. Journal of Power Sources 282, 265–276 (2015). doi:https://doi.org/10.1016/j.jpowsour.2015.01.072.

Ziemann, S., Müller, D.B., Schebek, L., Weil, M.: Modeling the potential impact of lithium recycling from EV batteries on lithium demand: A dynamic MFA approach. Resources, Conservation and Recycling 133, 76–85 (2018). doi:https://doi.org/10.1016/j.resconrec.2018.01.031.

Xiong, S., Ji, J., Ma, X., 2020. Environmental and economic evaluation of remanufacturing lithium-ion batteries from electric vehicles. Waste Manage. 102, 579–586.

Habib, K., Hamelin, L., Wenzel, H.: A dynamic perspective of the geopolitical supply risk of metals. Journal of Cleaner Production 133, 850–858 (2016). doi:https://doi.org/10.1016/j.jclepro.2016.05.118.

Olivetti, E.A., Ceder, G., Gaustad, G.G., Fu, X.: Lithium-Ion Battery Supply Chain Considerations: Analysis of Potential Bottlenecks in Critical Metals. Joule 1(2), 229–243 (2017). doi:https://doi.org/10.1016/j.joule.2017.08.019.

Zhu, X., Yu, L., 2019. Screening Contract Excitation Models Involving Closed-Loop Supply Chains Under Asymmetric Information Games: A Case Study with New Energy Vehicle Power Battery. Applied Sciences 9, 146.

Gu, X., Ieromonachou, P., Zhou, L., Tseng, M.-L., 2018a. Developing pricing strategy to optimise total profits in an electric vehicle battery closed loop supply chain. Journal of Cleaner Production 203, 376–385.

Gu, X.Y., Ieromonachou, P., Zhou, L., Tseng, M.L., 2018b. Optimising quantity of manufacturing and remanufacturing in an electric vehicle battery closed-loop supply chain. Ind. Manage. Data Syst. 118, 283–302.

Song, X.X., Hu, S.Y., Chen, D.J., Zhu, B., 2017. Estimation of Waste Battery Generation and Analysis of the Waste Battery Recycling System in China. Journal of Industrial Ecology 21, 57–69.

Qiao, Q., Zhao, F., Liu, Z., Hao, H., 2019. Electric vehicle recycling in China: Economic and environmental benefits. Resources, Conservation and Recycling 140, 45–53.

Dunn, J.B., Gaines, L., Sullivan, J., Wang, M.Q., 2012. Impact of Recycling on Cradle-to-Gate Energy Consumption and Greenhouse Gas Emissions of Automotive Lithium-Ion Batteries. Environmental Science & Technology 46, 12704–12710.

Ciez, R.E., Whitacre, J.F., 2019. Examining different recycling processes for lithium-ion batteries. Nature Sustainability 2, 148–156.

Dunn, J.B., Gaines, L., Kelly, J.C., James, C., Gallagher, K.G., 2015. The significance of Li-ion batteries in electric vehicle life-cycle energy and emissions and recycling's role in its reduction. Energy & Environmental Science 8, 158–168.

Clift, R., 2006. Sustainable development and its implications for chemical engineering. Chemical Engineering Science 61, 4179–4187.

ANL, 2018a. GREET. Argonne National Laboratory.

ANL, 2018b. BatPaC. Argonne National Laboratory, Argonne National Laboratory.

ANL, 2018c. ReCell, Argonne National Laboratory.

Linda, G., 2018. Lithium-ion battery recycling processes: Research towards a sustainable course. Sustainable Materials and Technologies.

Ciez, R.E., Whitacre, J.F., 2017. Comparison between cylindrical and prismatic lithium-ion cell costs using a process based cost model. Journal of Power Sources 340, 273–281.

Design and Utilization of Heritage Resources in Chongqing Air-Raid Shelter Project During the Anti-Japanese War

KaiGe Liu, DeXiang Deng, Xi Zhou, and JiaNi Liang

Abstract The heritage resources of the air-raid shelter project in Chongqing during the Anti-Japanese War have built the spiritual soul of the city, and should respond to its effective regeneration design and utilization, and lead the optimization of its air defense value, social value, cultural value, and commercial value with the continuation of the spirit. Make it fully integrated with environmental functions in the regeneration design, and enhance the environmental value of the area where it is located.

Keywords Air-raid shelter engineering heritage · Regeneration design and utilization · Spiritual soul · Value allocation · Environmental advantages · Value development

After the War of Resistance against Japan, Chongqing became the accompanying capital, and its special and important strategic position was bombarded by Japanese planes for several years. In the face of the Japanese bombing, a large number of underground air-raid shelters were built as shelters in Chongqing city, mainly to protect the lives of people in the city. After the victory of the Anti-Japanese War, the society gradually entered the era of peace. The function of the air-raid shelter during the war gradually faded and gradually faded out of people's sight. The air-raid shelter project during the Anti-Japanese War has become a cultural heritage of a special period.

At the end of the 1970s, the Third National Civil Air Defense Work Conference clearly stated that it would take the development path of "combination of peace and war," and the construction policy of civil air defense projects in the new era was established. Subsequently, the civil air defense department began to organize the construction of civil air defense projects that combined peace and war, including the comprehensive transformation of the original air defense projects, the establishment of large-scale air defense projects, and effective measures to explore the overall

K. Liu · D. Deng · X. Zhou (✉) · J. Liang
Chongqing University of Posts and Telecommunications, Chongqing, China
e-mail: 342862686@qq.com

R. G. Gutierrez (ed.), *Wastewater Technologies and Environmental Treatment*,
Springer Proceedings in Earth and Environmental Sciences,
https://doi.org/10.1007/978-3-030-61989-3_9

comprehensive benefits of air defense projects. At this time, the value of a large number of air-raid shelter engineering heritage resources left over from the Chongqing Anti-Japanese War began to appear, and the dormant Chongqing air-raid shelter engineering heritage during the Anti-Japanese War ushered in a revival of the environmental climate.

1 The Current Status of Regeneration and Utilization of the Heritage of the Air-Raid Shelter Project in Chongqing During the Anti-Japanese War

During the Anti-Japanese War, the total area of Chongqing's air-raid shelter project heritage resources was about 1.1 million square meters, mainly distributed in the current main urban area of Chongqing, among which Yuzhong District and Shapingba District were the most concentrated, Jiangbei District, Nan'an District, Jiulongpo District, and Dadukou District Also relatively more. Based on the establishment of the civil air defense route of China's "Combination of Peace and War", Chongqing has used some of the heritage resources of air-raid shelters during the Anti-Japanese War and transformed it into functions such as parking lots, building materials shopping malls, restaurants and other commercial purposes. In 2006, Chongqing suffered a severe drought and the government Forty-one air-raid shelters left over from the War of Resistance were opened in a concentrated manner so that citizens could escape the heat, and the public welfare value of air-raid shelters began to return. With the utilization of the heritage resources of the air-raid shelter project during the Anti-Japanese War, the value of the special heritage resource of the air-raid shelters re-entered people's vision.

According to relevant survey information, at this stage, only about 30% of air-raid shelter resources have been effectively used. However, most of the current air-raid shelters are relatively easy to use, and their strong value has not yet been displayed.

At present, the use of the heritage resources of the air-raid shelter project in Chongqing during the Anti-Japanese War has indeed achieved the effects of space utilization, income creation, and employment resolution. Generally speaking, there are some shortcomings.

1.1 Lack of Overall Rational Planning, Design and Utilization

In the protection of intangible cultural heritage, if the intangible cultural heritage cannot be reasonably developed, it will cause immeasurable losses to the place, the country, and even the entire human civilization (Chen 2013). During the

Anti-Japanese War in Chongqing, the air-raid shelters were widely distributed. There were great differences in size, height, depth, humidity, ventilation, altitude difference, and surrounding environment. The currently used parts are mostly those with large flow of people, convenient transportation, and high commercial value. Part of the air-raid shelters, most of them were not used. From an overall point of view, the use of air-raid shelter heritage resources in Chongqing during the War of Resistance against Japan lacked overall planning, design and utilization. It remained at the simple use of doorway surface decoration, random commercial use, low utilization rate, and did not exert its inherent air defense value, social value, and cultural value, the linkage configuration value of commercial value.

1.2 Lack of Emergency Preparedness

The main function of the air-raid shelter itself is a refuge facility during war (Deng et al. 2012), and the other functions are only the expansion of the air-raid shelter in peacetime. At present, most of the air-raid shelters used for commercial purposes in the air-raid shelters reviewed by the author cannot meet such conditions. Users of air-raid shelters used for commercial purposes will be locked at night to ensure that equipment and products are not lost. In the case of sudden war, the use of these commercial air-raid shelters lacks emergency preparedness. At the same time, these commercial air-raid shelters are full of commercial materials and equipment. In the event of sudden air strikes, the capacity of the air-raid shelters and the removal of equipment and materials are greatly reduced. The time difference will greatly reduce the prevention and control effect. These commercial air-raid shelters are difficult to quickly change and become a shelter for prevention and control when they truly respond to emergencies.

1.3 Lack of Historical and Cultural Transmission

At present, the use of Chongqing air-raid shelter heritage resources is mostly limited to profitable commercial transformation. Commercial profitability is a strong purpose, ignoring the excavation and transmission of the value of the anti-Japanese war history and culture of the air-raid shelter itself. Even if parts of the air-raid shelters are used for public welfare places to enjoy the coolness, they also ignore the value utilization of the air-raid shelters as a historical and cultural carrier of the War of Resistance.

1.4 Lack of Unique Value Discovery of Regional Environment

The reuse of air-raid shelter engineering heritage generally regards air-raid shelter as an ordinary cave space, ignoring the particularity of the regional environment and individual environment where the air-raid shelter is located, and lacks the excavation and utilization of its characteristic environmental value. Chongqing has the largest number of air-raid shelter engineering heritage resources during the Anti-Japanese War in China. The unique mountainous terrain and geological conditions of Chongqing make the internal environment of these air-raid shelter engineering heritage resources or the external environment, which is more advantageous than other cities. Create an important resource base for Chongqing's regional environmental characteristics. In fact, in Chongqing's existing reutilization of air-raid shelter engineering heritage, this environmental advantage value has not been exploited and explored.

2 Design and Utilization Strategy of Heritage Resources in Chongqing Air-Raid Shelter Project During the Anti-Japanese War

The heritage of the air-raid shelter project in Chongqing during the Anti-Japanese War is the actual relics left over from the National Anti-Japanese War. These historical relics are even non-renewable cultural resources, which demonstrate the tenacious resistance of the Chongqing people and the prosperity of the Chongqing companion era (Sun and Xue 2016). It is an integral part of the Anti-Japanese War culture. It bears the suffering and pain of the nation. The embodiment and symbol of the Chinese nation's self-improvement spirit. Therefore, the regeneration design and utilization of the air-raid shelter heritage resources in Chongqing during the Anti-Japanese War has great civil defense value, social value, cultural value and commercial value. At the same time, due to the unique topography and geological environment of Chongqing, the regional characteristic environmental value of the design and utilization of the heritage resources of these air-raid shelter projects is added.

2.1 Know the Basics, Design from the Original

First, the air-raid shelter is a product of war. As the war goes away, the concept of "preparing for war and famine" has faded and its original function is diluted. In peacetime, we still have an extremely close connection with the air-raid shelter (Jing 2012). Although the heritage resources of the air-raid shelter project in

Chongqing during the Anti-Japanese War can actively carry out the regeneration design and utilization of resources, they are designed for the national defense and civil defense functions. Therefore, the air-raid shelter should be retained, restored, and expanded during the regeneration design and utilization. The due air attack defense function, the modern defense and control capability, space capacity, ventilation system, etc. of its own structure, the emergency storage design of materials and communications in the air-raid shelter, and the modern transportation matching design of the air-raid shelter periphery are realized to realize its modern The upgrade of the prevention and control function ensures rapid prevention and control emergency and high-level air defense emergency in the modern war environment, and solves the rapid and safe evacuation of the masses. During the Anti-Japanese War, it was caused by the collapse of the air-raid shelter or too many people in the air-raid shelter. The tragedy of loss of precious life due to poor air circulation, high carbon dioxide gas content and poor breathing will not repeat itself in modern warfare.

Second, the air-raid shelter project in Chongqing during the Anti-Japanese War played a major role in preventing Japanese aircraft from bombing. It fundamentally consolidated its status as a companion capital, and broke the illusion that Japan wanted to start with the capital to solve the Chinese people's confidence in the War of Resistance. The air defense tragedies such as the eighteenth ladder at the entrance are also vivid. Therefore, the defense function of the air-raid shelter project in Chongqing during the Anti-Japanese War, the body's prevention and control function and the historical bearing, have important functions of national defense and civil defense education. The important point of the regeneration design and utilization of the heritage resources of the air-raid shelter project in Chongqing during the Anti-Japanese War is to help bring its own national defense education function and civil defense education function into play, improve national defense and civil defense awareness, and enhance the national defense and control, disaster resistance, and disaster reduction capabilities. History is not a dress rehearsal and cannot be repeated. How to make good use of the heritage resources of the Chongqing air-raid shelter project during the Anti-Japanese War to strengthen national defense education and civil defense education, and build a national and people's life and property safety line of defense is important for the design and utilization of the heritage resources of the Chongqing air-raid shelter project Point of view.

Third, the air-raid shelter heritage is the deep spiritual bond of the city, and it is the common experience of the city where all citizens live (Luo 2017). The air-raid shelter project in Chongqing during the Anti-Japanese War allowed the people of Chongqing to stand amidst the five-and-a-half years of indiscriminate bombing by the Japanese aircraft, and established a spiritual fortress for the people of the whole country. Its heritage resources carry the history of the Anti-Japanese War and the spiritual value of the Chinese Anti-Japanese War. The design and utilization of the heritage resources of the bomb shelter project in Chongqing during the Anti-Japanese War should promote this spiritual value, so that Chongqing people will use the unique spirit contained in the heritage resources of the bomb shelter during the Anti-Japanese War as a link to build an economic center and environmental ecological demonstration in the upper reaches of the Yangtze River in China; Let the people of the whole country, inspired by the unique spiritual value perception contained in the heritage resources of the air-raid shelters during the Anti-Japanese War, continuously activate ecological awareness, national spirit and patriotic enthusiasm, and empower the construction of ecological civilization, national prosperity and national rejuvenation dreams.

2.2 *Know the Characteristics of the Air-Raid Shelter, Develop and Design Based on the Characteristics*

During the Anti-Japanese War in Chongqing, the air-raid shelter engineering heritage resources were numerous and widely distributed. The geology, topography of most projects were different, and their structures and capacities were also very different. These constituted the unique characteristics and individualities of each air-raid shelter. Therefore, the design and utilization of the heritage resources of the air-raid shelter project in Chongqing during the Anti-Japanese War is to understand the respective characteristics and personalities of the external space and external environment of each air-defense, and carry out the design of differentiated individual forms, structures and functions to minimize the heritage of the air-raid shelter projects. Investment and optimization of recycling, while retaining the historical memory of wartime, so that it can play their own characteristic value in the personalized regeneration and utilization, highlighting the unique environmental characteristics of Chongqing.

2.3 *Know the Needs of the People Near the Air-Raid Shelter, and Design as Needed*

"Combination of peace and war" is our country's basic policy of maximizing the value utilization of civil air defense projects. Its core is the expansion of commercial and public welfare values led by emergency preparedness, and the expansion and use of non-air defense functions of civil air defense projects in peacetime. Economic and social benefits. The principle of combining peace and war has pointed out the direction for the regeneration design and utilization of the heritage resources of the air-raid shelter project in Chongqing during the Anti-Japanese War. However, we cannot rely on the previous single commercial leasing method for function expansion and benefit utilization. The specific needs of the market carry out its recycling design, that is, know what is needed and design on demand.

Knowing what you need and designing on demand is to understand the most urgent needs of users around the air-raid shelter, and plan and design the air-raid shelter according to the needs. For example, the Yuzhong Peninsula is densely populated, traffic is tight, and traffic jams have become the norm. The Yuzhong Peninsula will organically design and transform some of the air-raid shelters that were abandoned during the Anti-Japanese War into the vents or entrances and exits of rail transit line 1. These air-raid shelters have been rationally used. The social demand for heritage resources has greatly eased the problem of traffic tension. "Most of the existing air-raid shelters are located in the downtown area of various cities. These areas are often populated and demand for social welfare services are relatively strong. With the continuous development of modern cities, urban functional core areas are often under greater pressure in this regard. It is difficult to effectively meet the public's growing needs for public services" (Sun and Liu 2012). In this case, the air-raid shelter space is mainly designed for public leisure, attracting people from above-ground space to underground space, and the heritage resources of the air-raid shelter project in Chongqing during the Anti-Japanese War are cool in summer. The functional design is an excellent example of public welfare development for the performance of air-raid shelter heritage resources and the need for social summer refuge in stove cities. Most of the heritage resources of the air-raid shelter project in Chongqing during the Anti-Japanese War were distributed in the prosperous commercial districts, and its commercial value was very large. Since the Chongqing Municipal Government launched the air-raid shelter resource construction work, it can generate about 700 million yuan in revenue for the government every year, including providing corresponding support for a large number of social laborers. employment position.

Judging from the known use of the heritage of the air-raid shelter project in Chongqing during the Anti-Japanese War, the space for the regeneration of its heritage resources is very large. However, the corresponding needs of each air-raid shelter are not the same. We should fully investigate the social welfare needs or commercial needs of each air-raid shelter, and on the basis of giving priority to meeting the needs of social welfare, we should appropriately expand the needs of commercial services. Protective design development and reuse. Regardless of the expansion of public welfare value functions or the development and utilization of regeneration design for commercial value expansion, the basic functions of emergency prevention and control should be guaranteed to ensure that the citizens can recover the original air defense emergency function in time when the public is in an emergency. Play effectively.

2.4 Know Where the Air-Raid Shelter Is and Integrate It into the Regional Characteristic Environmental Design

There are three characteristics of Chongqing. One is the rich historical and cultural heritage contained in the wartime companion capital; the second is that the rugged and undulating mountainous terrain creates a good three-dimensional urban landscape; the third is the modern landscape of the city. These three characteristics are the direct reflection of Chongqing's historical and cultural environment, social environment and natural environment. Selectively protecting and repairing some old industrial buildings and supplementing with environmental renovation will surely make it another characteristic landscape in the city, giving full play to its special artistic value (Zhang 2009). The heritage resources of the air-raid shelter project in Chongqing during the Anti-Japanese War are about 1.1 million square meters. It is an "experiencer" of the special historical period of the Anti-Japanese War and a "witness" of the economic development of New China. It has rich social resources worthy of design and development. At the same time, these heritage resources are mainly distributed on the mountainsides, valleys, cliffs, and even the tops of the mountainous terrain in the main city of Chongqing. There are good space resources and natural resources worthy of full use.

Knowing the location of air-raid shelters and integrating them into environmental design is to fully understand the social environment and natural environment resources in which these air-raid shelter engineering heritage resources are located, so that the functions and styles of the regeneration design can be integrated into the local social or natural environment. Reuse these heritage resources with more forward-looking design thinking to make them glow with new vitality.

3 Conclusion

Chongqing, where the two rivers meet, is surrounded by mountains, and the terrain is uneven. The urban area does not even have a flat ground. The people of Chongqing have given full play to their wisdom and strength to build a unique multidimensional magical style three-dimensional dazzling and prosperous city. In the belly of such a city, the large and small air-raid shelter engineering heritage resources during the War of Resistance against Japan build up the spiritual soul of this city, respond to its effective regeneration design and use, and lead the air defense value, social value, cultural value, and commerce with the continuation of the spirit. The optimized configuration of value and environmental value makes it immortal in the regeneration design.

References

Jing, B.R.: Research on Chongqing Air-raid Shelter Construction during the Anti-Japanese War (2012).

Luo, J.: Reuse and diversification of air-raid shelters in urban core areas. Shandong Industrial Technology, 24.211 (2017).

Sun, J.Q., Liu, Y.: A Preliminary Study on the Protection Significance of Air-raid Shelters (System) in Chongqing. Anhui Architecture. 03 (2012).

Deng, S.Y., Wu, H., Wang, X.: Research on the Comprehensive Utilization of the Entrance Space of the Mountain City Air-raid Shelter——Taking Chongqing Dongshuimen Air-raid Shelter Entrance and External Space Design as an Example. Interior Design. 26.27 (2012).

Sun, J.Q., Xue, P.P.: Research on the Protection and Reuse of the Architectural Heritage of Chongqing Embassy and Consulate. Urban Development Research (02),-12(2016).
Zhang, D.J.: Analysis on the Protection and Utilization of Chongqing's Old Industrial Building Heritage. Interior Design (02), 6–11(2009).
Chen, Z.F.: Research on the Evaluation of the Development Value of Intangible Cultural Heritage Tourism Resources. Journal of Yangtze University (Social Science Edition). 12 (2013).

Design of Express Packaging Recycling System Based on Sustainable Concept

Xi Li, Xi Zhou, DeXiang Deng, and MengNan Wang

Abstract In recent years, global warming, excessive consumption of resources, and pollution of air and water resources have threatened the survival and development of human society. The rise of online shopping in China has stimulated the utilization rate of express packaging, and at the same time has resulted in the waste of express packaging, the phenomenon of faults in the recovery system consumes a lot of resources and runs counter to green economy and green development. This article is based on the concept of sustainable development, starting from the entire system of express packaging recycling, combining consumer needs, transportation processes, and corporate incentives, and discussing the "interactive ritual chain" of express packaging recycling based on the "green treasure box" Design the system.

Keywords Sustainable concept · Express packaging · Recycling · Interactive ceremony chain introduction

1 Introduction

In recent years, the academic community has reached the following consensus on the theory of sustainable development: resource protection and social development are two complementary and inseparable aspects, and nature protection should run through the framework of the entire social development, namely the population, resources and environment. Coordinated development (Lu 2010). At present, many countries have gradually transformed sustainable concepts into concrete and feasible sustainable design schemes, which have made great contributions to saving resources and protecting the environment. This proves that the design can indeed provide power and support for sustainable development (Carlo et al. 2014).

The packaging industry is an industry closely related to environmental protection and sustainable development, but at the same time, the high consumption of

X. Li · X. Zhou (✉) · D. Deng · M. Wang
Chongqing University of Posts and Telecommunications, Chongqing, China
e-mail: 342862686@qq.com

R. G. Gutierrez (ed.), *Wastewater Technologies and Environmental Treatment*, Springer Proceedings in Earth and Environmental Sciences,
https://doi.org/10.1007/978-3-030-61989-3_10

resources in the packaging industry and environmental pollution caused by packaging waste have become increasingly serious (Zhao and Ding 2011). With the rise of the express delivery industry in my country, a series of problems such as waste of resources and environmental pollution caused by improper recycling of express packaging have arisen. In February 2020, China's express delivery development index was 157.3, an increase of 31.6% year-on-year. In May, my country's express delivery totaled 1.102 billion parcels. The express delivery market still maintained a good development trend during the epidemic. However, according to data, the express delivery industry in China consumes 20.7 billion express waybills, about 3.1 billion woven bags, about 8.268 billion plastic bags, about 3.105 billion envelopes, about 9.922 billion boxes, and about 16.985 billion meters of tape inside. There are approximately 2.997 billion buffers (State Post Bureau 2016). Waste of express packaging resources has become a major environmental and social problem. Therefore, my country's express delivery industry urgently needs to build a mature express packaging recycling closed-loop system, so as to contribute to the protection of the ecological environment and promote the healthy and green development of the express delivery industry.

2 Iterative Upgrade of Express Packaging Under Sustainable Concept

Many countries have issued corresponding policies, regulations and technical standards for the recycling of products and logistics packaging. The United States began to pay attention to green packaging in the 1990s. The US government exempted enterprises from related taxes based on the enthusiasm of packaging recycling of various enterprises, and also explored packaging materials. For example, wine packaging is made of straw, which can be reused and is convenient for logistics and transportation. Germany's packaging waste management measures are to set recycling standards and time limits for different packaging wastes, and require transportation, agents, wholesalers, and retail companies to be responsible for the recycling of packaging materials. My country has also issued many policies to restrict and regulate the green development of the express delivery industry. At present, the total output value of my country's packaging industry accounts for more than 2% of the total social output value, ranking 14th among the 40 major industries in the national economy, and is known as the sunrise industry (Jin and Zhu 2011). In 2018, the State Council issued the "Opinions on Promoting the Coordinated Development of E-commerce and Express Logistics", encouraging e-commerce companies and express logistics companies to carry out supply chain green process reengineering, improving resource reuse rate and reducing corporate costs. Promote sustainable packaging technologies and materials, encourage e-commerce platforms to carry out green consumption activities and other relevant suggestions. In response, my country's express delivery industry has successively introduced new measures for the recycling of express packaging.

For example, ZTO Express chose to use separate billing for express delivery and packaging fees at express outlets to encourage consumers to pay attention to packaging costs; YTO Express tried to use chip eco-friendly bags to reduce the waste of packaging bags, implant chips in every bag, and track Record the source of the bag; STO Express uses recyclable environmental protection bags, which can be reused for 4–6 months; SF Express has launched the self-developed "Feng. BOX" recycling box, pioneering the use of zippers instead of sealing; launched The electronic waybill, which is easy to open and fold, has also changed the printing requirements, changing full-page printing to bottomless printing; using inflatable gourd film and other methods to save carbon emissions.

The above measures have achieved certain results, but they have not formed a collective response at the social level. Express packaging material providers and express transportation companies are all innovating in their own fields, but each field has its own way and has not formed a macro-integrated express packaging recycling design system.

3 The Complexity of Express Packaging Recycling System

System refers to a unified organic whole composed of several elements that are connected and interacting in a certain way (Zhou 2020). Express packaging recycling can be regarded as a system. The good operation of this system requires each link to interact and interact, so as to achieve the purpose of express packaging recycling. However, the recycling of the system involves many links and touch points, and the problems in each link may affect the virtuous circle of the system. Therefore, systematic design thinking is required for overall planning. At present, there are two main issues affecting recycling.

3.1 Packaging Materials in the Express Delivery Industry Are not Uniform

Express packaging is classified into paper, plastic, and wood based on materials; it can be divided into waybills, tapes, foam fillings, etc. according to usage. Most of the express delivery of plastic packaging will be thrown into the trash can and sent to the landfill together with other garbage. The time for degrading waste plastic is 400–500 years (Deng et al. 2011). Due to the large number of express delivery industries in my country, the express packaging materials used are also different. In foreign countries, Germany requires packaging producers and manufacturers to be fully responsible for product packaging, requires companies to bear corresponding responsibilities, and issues related policies such as "Waste Classification Instructions" and "Waste Cleaning Schedule" for companies to urge companies to establish a standardized system; The US government will grant corresponding tax reductions and exemptions to recycling green packaging companies. These measures can mobilize the enthusiasm of companies. Although I have promulgated policies related to garbage sorting, and various localities have responded positively, the lack of constraints and incentive policies for express packaging companies has caused companies to choose non-recyclable materials for packaging in order to reduce production costs, which is not conducive to promoting the reuse of express packaging.

3.2 *Consumers Lack Enthusiasm for Packaging Recycling*

China's waste recycling industry is market-centric (Kao et al. 2020). The recycling of express packaging requires a lot of personnel, funds and materials, and it is difficult for green enterprises to recycle. The main reason for this phenomenon is that consumers have not developed the habit of recycling, and they usually discard the packaging as garbage. Behavior shapes habits and habits change the environment. There are five levels in Maslow's needs. When users buy or send express delivery, they have already met the first level, but have not reached the second and third levels of security and emotion. Therefore, express packaging is easy to be ignored thrown away. Most of the online shopping user groups are young people. Few of them have the habit of hoarding discarded express packages in exchange for certain material rewards, and their response to material rewards is gradually weakening. But they pay more attention to time and value acquisition on a spiritual level.

The reuse of express packaging starts from changing people's consciousness and behavior habits in order to actively promote the implementation of the concept of sustainability in express packaging. Fogg, a professor at Stanford University, believes that three elements are essential to make people act. First, sufficient motivation; second, the ability to complete this behavior; third, the trigger for people to act (Eyal and Hoover 2017). After disassembling the express package, people did not put the express package in the recycling station or the express station, but chose to discard it in the first time. This is a manifestation of insufficient motivation. It is also possible that the user has this motive, but none of the behavior has occurred, which means that the trigger that motivates the consumer to act has not occurred.

4 Recycling System Design of "Interactive Ceremony Chain" for Express Packaging

As an environmentally responsible behavior that is easy to implement and enforceable, recycling behavior has gradually become a topic of global concern (Huffman et al. 2014). The environmental responsibility of express delivery is very important for predicting recycling behavior, and individual behavior intentions are also restricted by other individuals or groups (Gärling et al. 2003). The design of the express packaging reuse system is an innovation of the express packaging itself. It must also establish an interactive relationship between users and the express industry, express packaging suppliers and e-commerce platforms, increase the interest of consumers in using express packaging, and strengthen the relationship between the express industry and express packaging suppliers is connected, and an online and offline integrated interactive ceremony mechanism between environmental recycling companies and express packaging suppliers is established to achieve specific behavior transformation (Fig. 1).

Promoting the online business interaction model can further promote the enthusiasm of express packaging suppliers and green recycling companies. In real life, green recycling companies and packaging suppliers are two separate fields, each responsible for their respective functions. However, in the context of the smart era, online and offline cooperation has become the trend of future development. Packaging suppliers not only need to cooperate offline with various merchants, but also launch online innovation activities to build their own brands. Courier packaging suppliers can establish an offline interactive relationship with courier packaging and recycling companies. It is necessary to strengthen the recycling of each box and give each recycling box an "identity certificate", such as printing a scannable QR code to facilitate the enterprise uniformly classifies the recycled express packaging, selects the packaging that can be reused for processing, and places it on the market. At the same time, e-commerce has a huge demand for express packaging, and the cost will also be considered when buying express packaging boxes, but if e-commerce also joins the "interactive ceremony chain", a closed loop is formed with express packaging suppliers and recycling companies the circulatory system is the best choice in terms of development prospects.

Users are the largest group to return express packaging to green recycling companies. The "interactive ceremony chain" requires users to promote. Alipay is one of the most widely accepted software. Therefore, you can establish an "interactive ceremony chain" to use users as Mainly cut through the entrance,

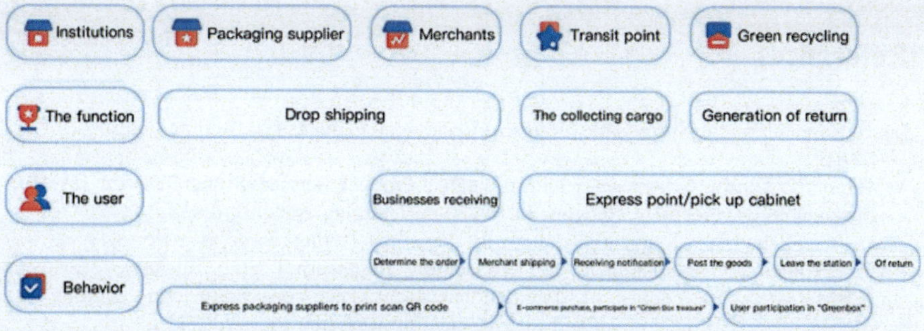

Fig. 1 Express packaging recycling system diagram

e-commerce, express delivery industry and express packaging suppliers launch courier boxes can be used to express courier freight and other activities. Taking into account the younger group's stronger emotional and spiritual needs, the event plans to launch the "Green Box Treasure" function: to advance in a way of emotional sustenance to increase the user viscosity of the product. Specifically, the user places the unused express packaging in a courier point or self-lifting cabinet, and after the placement is successful, scans the QR code on the packaging carton, which can increase the energy, role, and growth value of "Green Box Treasure". This energy can be used for freight collateral when accumulated to a certain degree; friends can also give each other the role of "Green Box Treasure". While increasing the fun of courier packaging recycling, users can also get certain benefits. To motivate users to participate in the activities of returning express packaging. For companies, save the cost of recycling express packaging. Therefore, an "interactive ceremony chain" based on the "green treasure box" was established, bringing together the interests of all parties, realizing information interaction and interaction with users, and achieving a solid recycling system cycle.

5 Conclusion

The rise of the express delivery industry consumes more and more environmental resources, and the implementation of the concept of sustainability in the express delivery industry is crucial to the development of human society. At present, the industry regulations and industry specifications of my country's express packaging recycling system still need to be improved, and the recycling of express packaging can hardly be advanced. This article discusses express packaging to establish an "interactive ritual chain" based on the "green box", starting from the material and emotional needs of consumers, and urging consumers to change their behavior patterns through design, so as to achieve partial promotion The virtuous cycle mode of the entire system operation has a positive impact on the protection of environmental resources.

References

Lu, S.:Research on the Sustainable Design of Residential Gardens [D]. Beijing Forestry University (2010).

V. Carlo, K. Cindy, S. Amulet, etc.:Sustainable Product Service System Design [M]. LeNS Sustainable Development Network International Learning Network (2014).

Zhao, Y.c., Ding, Y.: Recycling strategy of paper packaging waste in reverse logistics [C]. Proceedings of International Conference on Engineering and Business Management (EBM2011). Proceedings of International Conference on Engineering and Business Management (EBM2011). United States Scientific Research Press (2011).

State Post Bureau:Report on the development status and trend of green packaging in China's express delivery sector [J]. Green Packaging, (9) 9–10 (2016).

Jin, H.Y., Zhu, Z.W.:Environmental education research in packaging engineering education [C]. Education and Education Management (EEM 2011 V2). Intelligent Information Technology Application Society, 274–277 (2011).

Zhou, X.: 4D system view of product sustainable design [J/OL]. Packaging Engineering: 1–7 (2020).

Deng, Y., Lei, K., An, L., et al: Countermeasures on Control of Plastic Litter and Microplastic Pollution. 3E3S Web of Conferences 165, 02011 (2020) https://doi.org/10.1051/e3sconf/202016502011 CAES 2020 Bull Chin Acad Sci, vol. 33, pp. 1042–1051 (2018).

Kao, C.K., Sun, M.., Tang, Z.B., Luo, Y.P., Zeng, F.: Green Manufacturing Strategy of E-Commerce Express Packaging in China[J]. The International Technology Management Review.vol. 9, pp. 11–26 (2020).

Neil Eyal, Ryan Hoover, Addiction: Four Product Logics for Users to Form Habits [M]. CITIC Press (2017).

Huffman, AH; Werff, BRVD; Henning, JB; Watrous-Rodriguez, K.: When do recycling attitudes predict recycling? An investigation of self-reported versus observed behavior. [J]. Environ. Psychol. Vol, 38. pp,262–270 (2014).

Gärling, T.; Fujii, S.; Gärling, A.; Jakobsson, C.: Moderating effects of social value orientation on determinants of proenvironmental behavior intention. Environ. Psychol. Vol 23, pp 1–9 (2003).

Application of Sediment Fluidization by High Frequent Mechanical Vibrations into Slurry Transport

Wenkai Wang, Guoliang Yu, and Minxi Zhang

Abstract Friction reduction of sediment-transporting pipeline is very important in dredging engineering. Because of the two rheological behaviors of cohesive sediments, shear-thinning and thixotropy, cohesive sediments can usually be fluidized by the shear load which enables easier transportation of cohesive sediments and produces greater mobility of the sedimentary bed. This paper investigated the effect of the application of sediment fluidization by high frequent mechanical vibrations on the viscosity and transportation property of sludge and slurry using the method of settling ball and capillary rheometer respectively. The experimental results indicate that the high frequent mechanical vibrations can greatly influence the viscosity and the transport velocity of slurry. The vibrations applied on the sludge induce the mutual collision of the sediment particles which can break the bond between particles, so that the fine sediment bulk or sludge of high viscosity becomes a kind of slurry fluid which has quite low viscosity and good flow ability. Moreover, the effect of the vibration on the sediments decreases with the increasing distance to the vibration source, i.e., the closer the sediments to the vibration source, the lower the viscosity are. Meanwhile, the slurry velocity increases with the vibration duration until it reaches a stable value. Finally, two applications of the technique of sediment liquefaction are introduced in shortly.

Keywords Sediment-transporting · Pipeline · Cohesive sediments · Mechanical vibrations · Sediment fluidization

W. Wang
China Waterborne Transport Research Institute, Beijing 100088, China

G. Yu · M. Zhang (✉)
SKLOE, CISSE, School of Naval Architecture, Ocean and Civil Engineering,
Shanghai Jiao Tong University, Shanghai 200240, China
e-mail: zmx8888@sjtu.edu.cn

R. G. Gutierrez (ed.), *Wastewater Technologies and Environmental Treatment*,
Springer Proceedings in Earth and Environmental Sciences,
https://doi.org/10.1007/978-3-030-61989-3_11

1 Introduction

Sediment deposition widely occurs in Chinese rivers, lakes and other water bodies. In China, more than 50 thousand of rivers have its watershed area larger than 100 km^2, and 1500 rivers have its watershed area larger than 1000 km^2. According to a survey, 14 × 108 tons of sediment is annually deposited in reservoirs, rivers, canal channels, and lakes. 18 × 108 tons of sediment enters annually into estuaries and seas. Such large amount of sediment deposition causes a lot of problems in flood control, navigation, and water quality.

Dredge is one of the effective ways to remove the deposited sediment. To maintain the navigability of the harbors and waterways and sustain the storage capacity of existing reservoirs, periodical dredging activities are needed. In the last few years, along with the booming of dredging business, pipeline-transport of sludge has gradually become one of the main modes of transport for dredged materials in China due to its limited environmental impact, lower cost and higher efficiency. Many dredge projects such as the Liuhai dredge work in Beijing, the West Lake dredge work in Hangzhou, the dredge works of lakes in Xi' an and Xiamen, etc., utilized long pipelines to convey dredged material to designated sites. However, it is noted that as the pipeline becomes longer, more power would be required, and the easier jamming of pipelines would take place. Moreover, the sediment in most estuaries is fine-grained so that many of the world's ports are sited in muddy areas (Kirby 2010). The pipeline-transport are not very effective for cohesive sediments. The pipeline-transport of sludge becomes harder if deposits contain a high percentage of fine particles because the cohesiveness between such particles contributes to create resistant cohesive deposits, which are more difficult to remove than granular deposits (Mitchener and Torfs 1996). Due to the consolidation, cohesive sediments can "set up" and develop a hardened surface with time. The complex properties of cohesive sediments affect their erodibility which determines the efficiency of dredging. Hence, it is necessary to find out a more effective method to reduce the viscosity of slurry and the transport resistance in pipelines (Scott and Abt 2002).

Fine sediment plays a role in reduction of flow resistance in pipelines, which may be firstly proposed in China by Song et al. (1987). Zan and Wang (1991) indicated that the necessary condition of a low friction is that the coarse particles in the transported material had to be suspended and the friction could be reduced if certain amount of fine particles is added into the coarse particles. Similar findings were recorded by Matousek (2000). It is also found that the hydraulic gradient reduction effect resulting from phosphate clay added to sand and water slurries increased with sand concentration, decreased with particle size (Scott and Abt 2002).

The flowability of sediments, especially fine sediments, is closely related to its rheological properties (Zhang et al. 2017, 2019). Because of the two rheological behaviors of cohesive sediments, shear-thinning and thixotropy, cohesive sediments can usually be fluidized by the shear load which enables easier transportation of cohesive sediments and produces greater mobility of the sedimentary bed (Yang et al. 2014a). Two methods are generally used to improve the fluidization quality of cohesive materials. One is by external forces, such as vibration and the magnetic field (Chirone et al. 1993; Maring et al. 1994; Kuipers et al. 1996); the other is by altering the intrinsic properties of particles, e.g., by modifying surface characteristics and mixing with other particles having different sizes or shapes (Song et al. 2007).

In practical engineering, mechanical vibration is one of the most efficient methods for improving fluidization (Yang 2005), as the fluidization can be realized as vibration break bonds between particles in the bed (Tasirin 2001). A previous study by Yang et al. (2014b) found that high frequency vibration loads significantly reduce the viscosity of the cohesive sediments, the magnitude of the reduction increasing with the frequency of the vibration. In Dodaran et al. (2014), the dimension of flushing cone significantly increased due to vibrations. Thus, it can be assumed that the vibration can strongly improve the fluidization quality of cohesive sediments and increase their flowability.

To transport the raw sludge generally suffers large loss of head, and needs high energy consumption. The method of slurry production would directly affect the friction of sediment-transporting pipeline. In order to obtain a proper size distribution for slurry piping, Wang et al. (2005) discussed the effects of several modes of slurry production. Sun et al. (2001) proposed a method of vibrating the pipeline to increase transport efficiency and reduce energy consumption. The test of tailing transport conducted by Sun et al. has shown that the vibrating drag reduction may be practical and feasible. Based on their analysis of the reflection and transmission of stress wave, they regarded that the stress wave attenuated gradually as it propagated in the vertical direction of pipeline, forcing lower particles to move upward slightly to reduce the drag. However, it should be noted that this kind of vibration acting directly onto the pipelines may be harmful to the pipelines and reduce the lifespan of the pipelines.

In this study, the mechanical vibration is used to fluidize the sludge. The sludge herein means fine sediment deposited underwater, either in rivers, seas, oceans or lakes. The settling ball method is used to investigate the effect of the sludge fluidization induced by mechanical vibrations of high frequent and minute amplitudes on fluid properties. In addition a capillary rheometer is used to investigate the velocity of slurry flowing in the tube after being treated by vibration with different durations. Furthermore, the potential applications of the high frequent mechanical vibrations in dredging are introduced.

2 Theory

Fine sediment deposited underwater or sludge can be fluidized by high frequent and minute magnitude mechanical vibrations (Yang et al. 2014b; Zhang et al. 2019). When this kind of vibrations is applied on the deposited fine sediment or sludge with water mixture, the collisions of the sediment particle would take place in the manner of random and high speed. The vibrations of sediment particles would rapidly break the bonding forces, such as electric force, coherent force, etc., and enable the sediment particles in a very short moment to become independent and incoherent particles or molecules, so that the fine sediment bulk or sludge would become a kind of slurry fluid with good flowability, and then the original high viscosity would be significantly reduced.

3 Experiments

The results of sediment fluidization can be qualified in terms of viscosity coefficient. The more the viscosity of the sludge is reduced after the treatment by the mechanical vibration method, the better the sludge is fluidized. In this study, the viscosity of slurry was measured with the help of the settling balls and the flowability of the slurry was measured by a capillary rheometer.

3.1 Experimental Setup

1. The method of settling ball:

The small settling ball was a steel hollow-core ball with a density of 3100 kg/m^3. The ball was hung on a wheel and connected to a small marker by a tiny thread. The uplift velocity of the marker on the thread was the speed of the ball settling into the slurry. The settling process was recorded by a movie camera SR100E with a frame speed of 24 frames per second.

The mechanical vibrations were executed by a vibration bar which was 5 cm in diameter and 50 cm in length. The design vibration magnitudes were about 0.5 mm and the vibration frequency was 200 Hz. Its vibration power was 1.4 KW.

The experiment slurry basin was illustrated in Fig. 1. The settling velocity was measured at six different locations B1–B6, i.e., 4, 6, 8, 10, 12 and 14 cm from the right rim of the vibration bar.

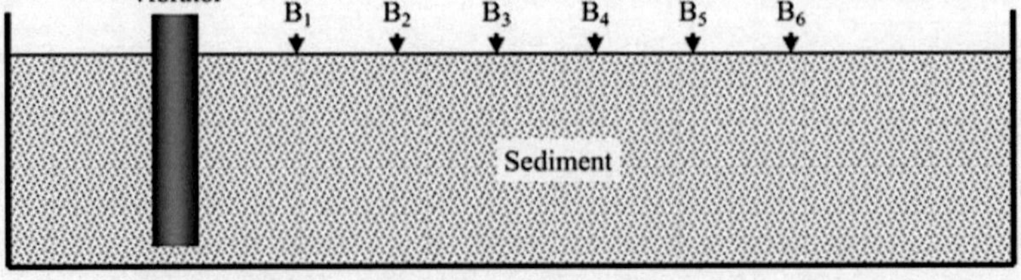

Fig. 1 The experimental slurry basin

According to the experiment observation, the settling of the small ball was very slow and reached a steady speed very fast. Based on the forces balance on the settling ball, the settling velocity is written as

$$V = \frac{(\rho' - \rho)gr^2}{72\eta} \tag{1}$$

where, V is the settling velocity, ρ' is the density of the ball, ρ is the density of the ball, r is the ball radius, η is the viscosity, g is the gravity acceleration.

2. The method of capillary rheometer:

The experiment equipment system was composed of a container, a capillary tube, a collector, and a support table as shown in Fig. 2. The capillary tube was a steel pipe with a diameter of 0.7 cm. The cylindrical container was 50 cm high and 30 cm in diameter, used to contain experimental slurry. Its bottom was connected with the tube. The collector was used for receiving the outflow slurry from the tube. The receiving process was recorded by a movie camera in order to precisely determine the receiving duration.

3.2 *Experimental Materials*

Two different types of sludge are used in this study, which are taken from the canal of Shanghai Jiao Tong University (sludge 1) and the estuary of the Yangtze river (sludge 2). The original density is 1830 kg/m^3 and 1960 kg/m^3, respectively.

In the settling balls test, these two types of sludge are divided evenly into three parts and put into 6 slurry basins, respectively. Before beginning the tests, the balls are buried into the sludge just below the mud surface and stand for days (1, 3 and 7) to let the disturbed sludges consolidate. Thus, the sludges 1 is divided to 3 parts

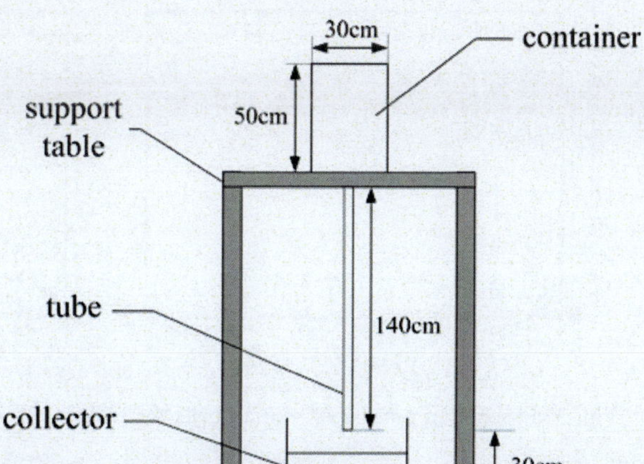

Fig. 2 Sketch of capillary rheometer

which are named as sample 1 (1 day), sample 2 (3 days) and sample 3 (7 days). Similarly, the sludge 2 is divide to sample 4, 5 and 6.

In the capillary rheometer test, the sludges were elaborated with various water contents that allowed the samples to possess different values of concentrations. The sludge 2, which is made to 9 samples with different wet bulk densities, are used to investigate the effect of the concentration of the slurry on the enhancement of the flowability. The sludge 1, which is made to 4 samples with different wet bulk densities, are used to investigate the effect of the vibration time on the enhancement of the flowability.

4 Results and Analysis

4.1 Influence of Vibration on Viscosity

The settling velocity and the penetration depth of the balls are measured by the marker. For example, Fig. 3 illustrates the penetration process of settling ball for sample 1 and 4. The friction of the wheel can be ignored. Because of the existence of the thread and the marker, the steady settling velocity is written as

$$V = \frac{(\rho' - \rho_1 - \rho)gr^2}{72\eta} \tag{2}$$

where, ρ_1 is the equivalent density, i.e., the total weight of the thread and the marker divided by the volume of the ball, which was 190 kg/m^3. Hence, the viscosity of the slurry can be calculated by

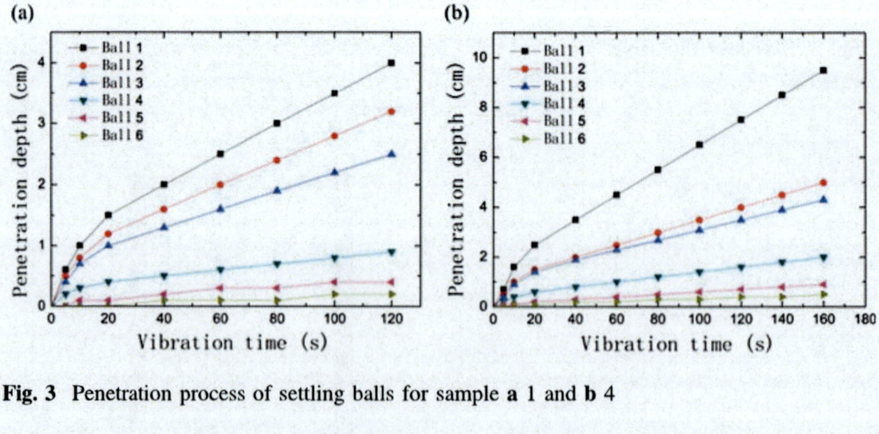

Fig. 3 Penetration process of settling balls for sample **a** 1 and **b** 4

$$\eta = \frac{(\rho' - \rho_1 - \rho)gr^2}{72V} \tag{3}$$

The corresponding viscosities at six locations were calculated. The result indicates that the high frequent mechanical vibrations can greatly influence the viscosity. Figure 4 illustrates the percentage of viscosity reduction of the sludges at different locations after being treated by the vibration. The result indicates that the high frequent mechanical vibrations can greatly influence the viscosity. For the sludges which are near the vibration bar, the viscosities can be reduced by over 90%. The vibrations applied on the sludge induce the mutual collision of the sediment particles which can break the bond between particles, so that the fine sediment bulk or sludge of high viscosity becomes a kind of slurry fluid which has quite low viscosity. Moreover, as shown in Fig. 4, the effect of the vibration on the sediments decreases with the increasing distance to the vibration source, i.e., the closer the sediments to the vibration source, the lower the viscosity are. It is interesting that the influence ranges of the vibration are different for different types of sludges. For sludge 1 (sample 1, 2 and 3), the reduction percentage of viscosity is over 80% in the range of 8 cm. Then, the influence of vibration becomes weak rapidly, for sample 3, when the distance to the vibration center is 14 cm the reduction percentage is only about 13%. However, for sludge 2 (sample 4, 5 and 6), the critical distance is 10 cm.

4.2 *Influence on Vibration on Flowability*

The velocities of original sludge and vibration treated sludge flowing in the capillary tube are listed in Table 1. The tested sludge was taken from the estuary of the Yangtze river (sludge 2). It is clearly shown in Table 1 that the slurry velocity was

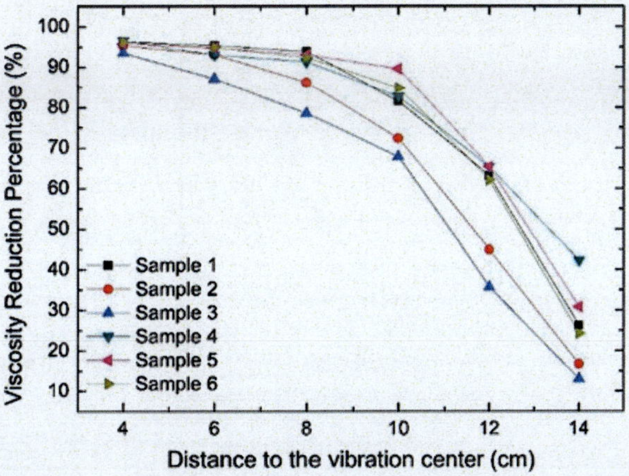

Fig. 4 Viscosity reduction after vibration treatment

Table 1 Velocity in capillary tube pre/post-vibration treatment

Density (g/cm^3)	V pre-treatment (cm/s)	V post-treatment (cm/s)	Velocity increment (%)
1.02	163.66	170.75	4.3
1.05	157.07	166.93	6.1
1.1	150.07	160.42	6.9
1.15	147.32	158.23	7.4
1.2	143.07	156.13	9.1
1.24	130.99	145.45	11
1.3	124.29	142.67	14.8
1.35	122.58	148.40	21.1

significantly increased after the vibration treatment. The denser the fluid, the more significant the increment would be. For the sludge with a sediment concentration of 23% (density 1350 kg/m^3), the slurry velocity was increased 21.1% after fluidization treatment, which indicated the sludge fluidization technique would have great significance to improve the sediment-transport efficiency and reduce the energy consumption as well.

The sludge 1, which is made to 4 samples with different wet bulk densities, are used to investigate the effect of the vibration time on the enhancement of the flowability. Figure 5 shows that the slurry velocity increases with the vibration duration for the second type of sludge with different concentration of sediment. The reason behind it is the thixotropy of the cohesive sediments. The thixotropy represents that the viscosity of the sediments is time-dependent, i.e., when subjected to a certain shear rate, the viscosity of the sediments decreases with the increasing

Fig. 5 Velocity increment of slurry against vibration treatment duration

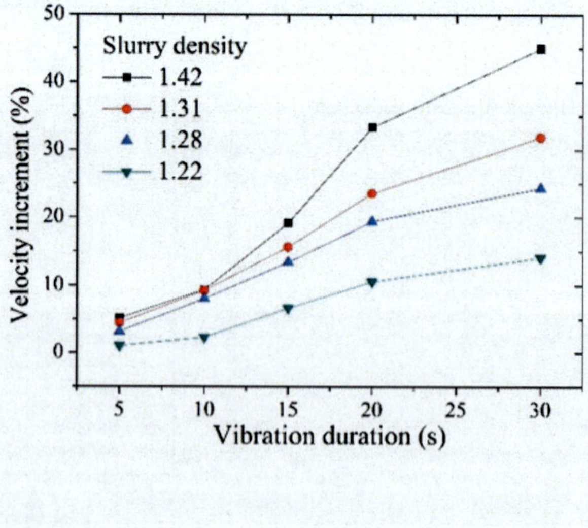

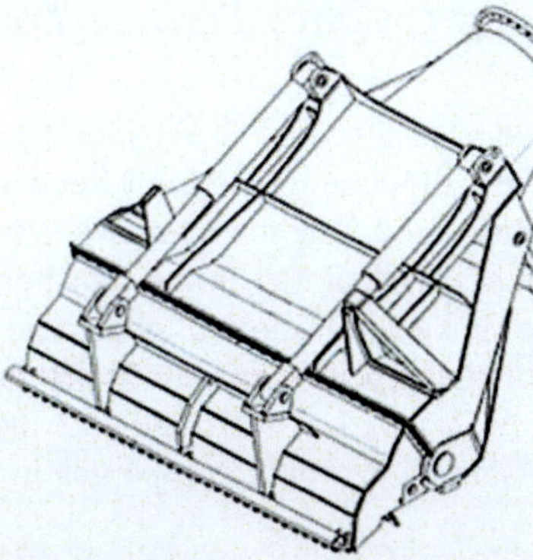

Fig. 6 Dredger head of vibration function

shear time (Yang et al. 2014a). Thus, the mechanical vibration can fluidize the sludge in very short time, the more the shear (or the vibration in this case) lasts in the time, the more the material is fluidized until the destruction and recovery of the internal structures of the sediments reaches a balance.

5 Applications

As aforesaid, the high frequent mechanical vibrations in the sludge can rapidly make the sludge fluidized, and can decrease its viscosity and improve its flowability as well. Hence, the limit deposit velocity can be more easily satisfied and the jamming phenomenon can be therefore more easily avoided. Accordingly, two conceptual designs for the application of high frequent vibration fluidization technique, patented with No. 200710172600.8 and 200510029894.8 respectively, are proposed as follows.

5.1 New Dredging Head

The high frequent vibration fluidization technique can be applied to dredging heads to loose the underwater sludge composed by fine sediment. As shown in Fig. 6, a roller with high frequent mechanical vibrations is installed on dredger heads. This roller continuously vibrates underwater fine sediment as it parades with the dredger head. The fluidized sediment in the rail of the roller is then easily sucked up by the suction pump of the dredger.

5.2 Long Distance Pipeline Transport

In the design of a sediment transporting pipeline, a slurry maker is built up at the origin of the pipeline to use high frequent mechanical vibrations to liquefy sludge before transported. The fine particles is then released from the bonding and act as a kind of lubricate material in the transporting pipeline.

As the slurry is transported longer, the more coarse particles would deposit to the bottom of the pipeline and the more lubrication function would be lost. Slurry makers are installed in the middle of the long pipeline acting as relay stations to keep the slurry's hydraulic properties in favor of transport.

6 Conclusions

In this study, the mechanical vibration is used to fluidize the sludge. The viscosity and flowability of the fluidized sludge is investigated in the laboratory. The following conclusions can be stated.

The high frequent mechanical vibrations can greatly influence the viscosity and transport velocity of slurry. The vibrations applied on the sludge induce the mutual collision of the sediment particles which can break the bond between particles, so that the fine sediment bulk or sludge of high viscosity becomes a kind of slurry fluid which has quite low viscosity and good flow ability. Moreover, the effect of the vibration on the sediments decreases with the increasing distance to the vibration source, i.e., the closer the sediments to the vibration source, the lower the viscosity are. The effect of the fluidization is much stronger for the slurry with higher strength. Meanwhile, the slurry velocity increases with the increasing vibration duration until it reaches a stable value. The high frequent vibration fluidization technique could be used in dredging engineering to pre-process the sludge such as a new dredging head or a slurry maker.

Acknowledgements The authors would like to extend grateful appreciation to the financial support from the National Natural Science Foundation of China (Grant No. 51779137), and the CCC National Engineering Research Center of Dredging Technology and Equipment Co. Ltd.

References

Chirone, R., Massimilla, L., and Russo, S.: Bubble-free fluidization of a cohesive powder in an acoustic field. Chemical Engineering Science, 48(1), 41–52 (1993).

Dodaran, A.A., Park, S.K., and Mardashti, A.: Experimental Investigation of the Effect of Vibrator Machine for Desilting Sediment in the Pressure Flushing. Journal of Scientific Research, 6(3), 475–485 (2014).

Kirby, R.: Minimising Harbour Siltation: The Economic and Environmental Future of Ports. Tri International Conference on Ports, 709–718 (2010).

Kuipers, N.J.M., Stamhuis, E.J., and Beenackers, A.A.C.M.: Fluidization of potato starch in a stirred vibrating fluidized bed. Chemical Engineering Science, 51 (11), 2727–2732 (1996).
Marring, E., Hoffmann, A.C., and Janssen, L.P.B.M.: The effect of vibration on the fluidization behaviour of some cohesive powders. Powder Technology., 79 (1), 1–10 (1994).
Matousek, V.: Experimental investigation of slurry flow mechanism in a pipeline. Journal of Dredging Engineering, 2(3): 18–39 (2000).
Mitchener, H., and Torfs, H.: Erosion of mud/sand mixtures. Coastal Engineering, 29(1), 1–25 (1996).
Scott, S. H., and Abt, S.R.: Hydraulic transport of fine and coarse sediment mixtures in pipelines. Journal of Transportation Engineering, 128, 1–8 (2002).
Song T.C., Nien N. and Wan Z.H.: Resistance of sediment transport in pipelines. Journal of Sediment Research, 1987, June, pp. 30–41 (in Chinese) (1987).
Song, L.Y., Zhou, T., Yang, J.S., Zeng, P., and Chen, G.Q.: Research on fluidization of fine powder with additive particles. China Powder Science Technology, 6, 30–35 (2007).
Sun, Y.Z., Wu, A.X.G, and Li, J.H.: Vibration Drag reduction of Pipeline Transport of High Density Slurry. Journal of Mining and Metallurgical Engineering, 04 (in Chinese) (2001).
Tasirin, S.M.: Fluidization behavior of vibrated and aerated beds of starch powders. Journal of Chemical Engineering of Japan, 34 (10), 1251–1258 (2001).
Wang, S.Z., Zheng, L.M., and Hou, D.K.: Discussion on the Method of Raising Sediment Piping Efficiency. Journal of South-to-North Water Diversion and Water Science & Technology, 2005, 04(in Chinese) (2005).
Yang, W. Yu, G., and Wang, H.: Rheological properties of dense natural cohesive sediments subject to shear loadings. International Journal of Sediment Research, 29 (4), 454–470 (2014a).
Yang, W., Tan, S.K., Wang, H., and Yu, G.: Rheological properties of bed sediments subjected to shear and vibration loads. Journal of Waterway, Port, Coastal, Ocean Engineering, https://doi.org/10.1061/(asce)ww.1943-5460.0000211 (2014b).
Yang, W.C.: Fluidization of fine cohesive powders and nanoparticles—a review. Journal of the Chinese Institute of Chemical Engineers, 36, 1–15 (2005).
Zan, H.H. and Wang, Z.N.: Interact of different size of particles in pipeline transported slurry and its friction reduction effect. Journal of Yunan metallurgy, 20(2), 33–36 (in Chinese) (1991).
Zhang, M., Yu, G., Rovere, A.L., and Ranzi, R.: Erodibility of fluidized cohesive sediments in unidirectional open flows. Ocean Eng., 130, 523–530 (2017).
Zhang, M., Yu, G., Zhu, W., and Li, M.: Experimental Study on the Angle of Repose of Submerged Cohesive Sediments. Journal of Waterway Port Coastal and Ocean Engineering-asce 145.3 (2019).